Birdsill Holly, of Lockport, N. Y., " Father " of the Central-Station Heating
Industry   *Frontispiece*

DISTRICT HEATING

A Brief Exposition of the Development of District Heating and its Position among Public Utilities

BY

S. MORGAN BUSHNELL

Past President, National District Heating Association
Member, The American Society of Heating and Ventilating Engineers
Member, American Institute of Electrical Engineers

AND

FRED. B. ORR

Commercial Engineer, Illinois Maintenance Company

Fredonia Books
Amsterdam, The Netherlands

District Heating:
A Brief Exposition of the Development of District
Heating and its Position Among Public Utilities

by
S. Morgan Bushnell
Fred B. Orr

ISBN: 1-4101-0229-7

Reprinted from the 1915 edition

Fredonia Books
Amsterdam, The Netherlands
http://www.fredoniabooks.com

PREFACE

THE preparation of this work was prompted by the hope that it might call attention to certain phases of District Heating which have not been extensively discussed by other writers. It has been the endeavor of the authors to take up in a brief manner the development of the art of District Heating, as far as possible, from a non-technical standpoint, in order that those who have not had experience with this subject might obtain a general knowledge of District Heating and its relation to other public utilities. As a rule the discussion of the more involved technical points has been avoided, the reader being referred to standard text-books and other engineering literature which take up these subjects in detail.

The object of the book is not only to impart a general knowledge of District Heating to those who wish to study the subject, but also to promote the interests of District Heating companies. For this purpose various points on economical operation are suggested, as illustrated in the methods practiced by companies which have made a success in this line of work. Comparative data are presented indicating the prices that can reasonably be charged for heating service.

The authors have availed themselves of the generous assistance of many men who are prominently identified directly or indirectly with heating companies and sincere appreciation is felt for the kindly interest shown in so many ways. This assistance has tended to make a volume which is less local and more comprehensive in character, reflecting the views of many, rather than of a few.

THE AUTHORS.

CHICAGO, ILL.
June, 1915.

TABLE OF CONTENTS

LIST OF ILLUSTRATIONS

DISTRICT HEATING

CHAPTER I

ORIGIN AND DEVELOPMENT OF DISTRICT HEATING

FROM the earliest dawn of history, mankind when dwelling in those regions situated outside the tropics, has found it necessary to obtain heat by means of fire. The origin of fire is lost in the mists of antiquity. Whether the first notion of fire was obtained from the blazing volcano, or from the forest fire, supposedly lit by the sun's rays, focussed through a piece of quartz or in some other way it is now impossible to determine.

Almost every country in the world has its myths and legends relating to the origin of fire. The Indians of North America claim that the first fire was started from the hoof of a great buffalo striking the flint-rock of the prairie, while the old Greeks had the legend of Prometheus carrying his lighted torch down to earth from the sun. The ancient savage tribes found that by briskly rubbing two sticks together they could develop the spark which would start a simple camp-fire.

In the castles of the middle ages great fire-places were constructed on which the logs were piled high on cold, frosty evenings, while the members of the household gathered around the open fire. Later stoves were provided for a more economical consumption of coal or wood and, still later, furnaces, from which hot-air pipes were led to various portions of the building. All of these applications of heat were local from their very nature.

The more recent methods of transmitting heat by means of steam and hot-water systems have brought what is thus far the last development in the evolution of heating systems, namely,

" District Heating." It is interesting to note that the development of district heating has been very closely related to the development of the central station industry for the distribution of electric light and power; in fact, many have supposed that the development of central heating was a result of the adoption of central lighting and power systems.

While it is true that certain systems of district heating, such as the Yaryan system, have been developed as a part of a system for the economical production of light and power, yet the fact remains that central station heating or district heating antedates the development of central station lighting by a number of years. It is doubtless true that in the early days of steam heating various people have heated more than one building from a single source. However, just as Thomas A. Edison is looked upon as the father of the central lighting station, so in the heating industry there is one man generally named as the pioneer inventor of central station heating, Mr. Birdsill Holly of Lockport, New York. Mr. Holly's inventive genius found an outlet not only in devising a practical system of central station heating, but also in the invention of the Silsby steam fire-engine, the rotary pump, and last, but not least, the Holly system of direct pressure water supply, which has been used in a great many cities all over the country.

THE BEGINNING OF DISTRICT HEATING

In 1876, Mr. Holly ran an underground line from a boiler in his residence to a barn at the rear of his property and later connected an adjoining house. In 1877, he constructed his first experimental plant at Lockport, in the State of New York, and a number of residences, stores and offices were successfully heated during the following winter.

The first installation consisted of about three miles of street mains and in 1878 another mile was added to the system. The installation was made in accordance with designs patented by Mr. Holly, his inventions applying particularly to the distribution of heat. It is a fact well-known by engineers that with a steam piping system it is necessary to allow for the expansion of piping due to the effect of heat, and up to the time of the Holly system, the only method of allowing for this expansion was by means of elbows in the piping. Mr. Holly designed the first expansion joint for taking care of the expansion and contraction of piping and his method of insulating the piping underground was not so very different from the more approved methods in use today.

At the time this system of heating was designed it attracted considerable attention throughout the country and a company was organized known as the Holly Steam Combination Company, Limited. This company acquired Mr. Holly's patents on the apparatus which he designed and also the patent on the system itself. It was not long before heating plants were being established at various points all over the country. An agent of the company would visit a certain town and perhaps call a meeting of some of the representative merchants and bankers of the locality to whom the plan would be unfolded. A company would then be formed and the stock placed with the leading residents of the town. A contract would be let for the installation of pipe-line and heating boilers and in a few months another district heating system would be under way.

FORMATION OF AMERICAN DISTRICT STEAM COMPANY

In 1882, the American District Steam Company was formed and purchased the patent rights of the Holly Steam Combination Co., Ltd., also the other inventions which Mr. Holly had brought out in the previous five years. This company, which, by the way, is considerably older than the General Electric Company, has spent a great deal of time and money in experimenting and in designing central station heating systems, and has contributed in no small degree to the development of this line of business. It is true that a great many of the original plants were financially unsuccessful, just as a majority of the first lighting central stations were unsuccessful; however, in spite of all these dis-discouragements and draw-backs, the men interested in the district heating business have gone steadily on and today the outlook for district heating is brighter than at any period of its history. It is interesting to note that while the first central station for heating was established in 1877, it was not until October 29, 1879, that Thomas A. Edison burned a carbonized piece of cotton sewing thread (in vacuum) for forty hours and demonstrated the possibility of the incandescent lamp. Two years later, after working actively on the various phases of the lighting problem, Edison was able to exhibit at the Paris exposition, a complete system of electric lighting, including the dynamo, wiring and lamps. In 1882, the Pearl Street station of the New York Edison Company was begun, this being the first central station of any importance for the production and manufacture of electricity for lighting purposes.

Turning to the subject of district heating in New York, we find the start had been made several years previously. In the winter of 1879–80, Mr. Cassius C. Peck went to New York and assisted Mr. Wallace G. Andrews, of Cleveland, Ohio, in securing a franchise for the New York Steam Company. The following year this company was consolidated with another company which did not have a franchise, the pipes were laid and the first customer was a printing establishment to which heat was supplied in May, 1882. It was not until 1883 that steam was supplied to customers in considerable quantities. From that time on, the business grew rapidly from year to year and, as a result, the New York Steam Company has always been the largest and most important company in the country and probably in the world for the distribution and sale of steam.

FIRST DISTRICT HEATING PLANTS USED LIVE STEAM

In connection with the early history of district heating, it should be noted that the first steam heating plants used a direct steam heating system or live steam instead of exhaust steam from a power plant. In fact, probably the first electric company that became interested to any extent in central station heating was the Brush Swan Electric Light Company, of Auburn, New York, which installed, in 1885, a direct steam heating plant. However, another proof of Mr. Holly's far-sighted grasp of the situation is shown in the application for a patent, made out April 18, 1881, by him, covering in detail various methods of exhaust heating in use today and including the combination of direct low pressure steam from boilers with exhaust steam from engines, and high pressure feeders where desired from the same system of boilers. This patent, which covers a most comprehensive and general theory of exhaust steam heating, was issued to Mr. Holly, June 27, 1882. For several years, the value of this patent seems to have been only partially appreciated and it was not until 1889 that any considerable system was installed, based on the use of exhaust steam.

In December, 1889, a system was installed by the Ottumwa Railway & Light Company in the city of Ottumwa, Iowa, consisting of nearly 6,000 ft. of mains. This system was designed for the purpose of heating from house to house by means of exhaust steam. The fact that as far back as 1878, direct steam heating systems were installed in Auburn, N. Y., Springfield, Mass., and Detroit, Mich., shows that it was fully ten years after the first general knowledge of district heating that exhaust steam systems were adopted.

In Detroit the heating business has been gradually extended and was for years conducted by two companies, the Edison Illuminating Company, of Detroit, and the Murphy Power Company. In the spring of 1914, a consolidation of the two companies was effected giving the Edison Illuminating Company of Detroit a total heating business of about 1,500 customers.

The Central Heating Company of Milwaukee is another company which has gone into the business of central heating on a very extensive scale, and with marked success. In the city of Indianapolis, there is a large amount of heating business. There were formerly three companies furnishing district heating, the Merchants' Heat and Light Company, the Indianapolis Light and Heat Company and the Home Heat and Light Company. The first two distributed heat by means of exhaust steam from the engines of their electric lighting stations, while the last used a hot-water system; the water for this system being largely heated by exhaust steam. Recently the plant of the Home Heat and Light Company was bought up by the Merchants' Heat & Light Company, leaving at present two companies in the field.

SMALLER TOWNS TAKE UP DISTRICT HEATING

There are district heating plants not only in the larger cities, but also in many of the smaller towns throughout the country. It has been found most satisfactory and economical, for a great many colleges and various institutions having a number of buildings, to have one central heating plant to supply all the various buildings.   On the Long farm near Kansas City, there are about two miles of heating mains, connecting various barns, houses for the employees, and the owner's residence.   Some of the finest blooded horses are kept on this farm and by distributing heat in this way to the different barns, the danger of fire is eliminated.   Ball Brothers, of Muncie, Ind., occupy seven houses connected with a system of mains about one-half mile in length, the central heating plant being located on a switch track, thus reducing the cost of fuel.   However, it is unnecessary to enumerate the many cities and towns all over the country in which district heating has been adopted as a matter of convenience and economy.

DEVELOPMENT OF COMPANIES FORMED TO OPERATE EXISTING PLANTS

Up to the year 1900, the development of district heating was based on furnishing the service from the main central source of supply, using a system of distributing pipes, running through the streets. About that time, there commenced a development of district heating along slightly different lines. Most of the large electric light and power stations began to feel a demand from their larger customers for a company that would complete the cycle of their requirements and enable building

owners to contract for the three forms of energy required for
the operation of a building, viz.: light, heat and mechanical
power. In order to supply this want, and at the same time to
avoid the enormous expense connected with the installation of
steam pipes in the congested districts of a large city, a company
was organized in Chicago in 1900, known as the Illinois Maintenance Company.

The charter of this company was a broad one, but the main
object of the company has always been to operate heating plants.
The company was operating on a scheme somewhat diverse
from that employed by steam heating companies up to that
time. It installed no plants of its own, and, except in one or
two cases, installed no boilers. Its plan was simply to contract
to operate boiler plants, already installed in buildings so as to
furnish heat and steam for the building and for other buildings.
This scheme at first sight may seem a radical departure, and
it might be thought that central station heating was being lost
sight of, but a careful study of the situation will prove the reverse
to be true.

In the first place, the conditions to be met with in a large
city are radically different from those in small towns, having
a house to house business, and also having the advantage of a
marked difference in economy between the small heating boiler
used in houses and the large one of 300 to 500 h.p. found in the
central station. Careful tests which have been made on residence boilers under test conditions show an average efficiency
of 40 to 50%. It is probable that under ordinary working conditions the percentage might even be less, perhaps 35%. The
large boilers used in central stations will give an efficiency of
between 70 and 80%. If we would assume an average efficiency
on the large boilers of 65% we would have right there a difference of nearly 50% between the coal consumption for the same
amount of heat in the smaller boiler and that required with the
large boiler. Furthermore, the small house heating boiler uses
anthracite coal which costs fully 100% more than the soft coal
used in the larger boiler installations. Accordingly, if it were
not for transmission losses, a central heating plant could supply

a given amount of heat with about one-fourth the coal cost of a small house heating boiler.

THE NEW YORK STEAM CO.
DISTRICT "B"

FIG. 1.—Downtown-heating system of the New York Steam Co.

In the center of large cities like New York and Chicago, this difference disappears, as the ordinary business block will require from 100 to 500 h.p. in boiler capacity and the difference in efficiency between the 100 and 500 h.p. boiler is comparatively small. On the other hand, the difficulty of installing elaborate systems of street mains in city streets, already crowded and con-

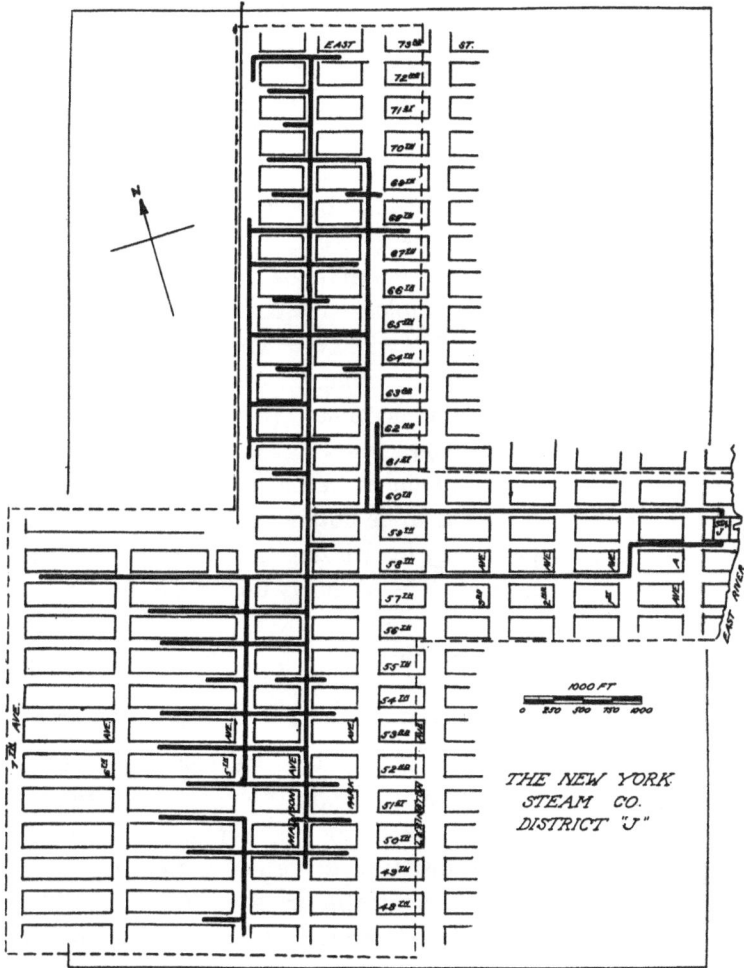

FIG. 2.—Uptown-heating system of the New York Steam Co.

gested with the pipes of various other utilities, is almost insuperable. Accordingly the Chicago company adopted a consistent plan of operation which used boilers already installed in buildings and also connected by pipe lines, adjacent buildings in order to

gain economy by shutting down the small plants and operating
only the larger ones.   This, to be sure, requires occasional
crossing of streets, but does not require anything like the expense
which would be involved in a comprehensive system.

Soon after this system was started, the Boston Edison Company inaugurated in Boston a similar system, operated under
the direct management of the company.   A little later the New
York Service Company was organized and operates to a limited
extent in the business district of New York.   In Philadelphia
and St. Louis a similar scheme has been worked out to a certain
extent.   While in all of these different cities, there are individual characteristics and slight variations from the general plan,
the net result has been a rapid increase in the field of district
heating.   In New York City the New York Steam Company
has improved its load factor by using its battery of boilers for the
manufacture of ice in the summer time.   In other words, the same
generating plant that produces power for heat in the winter time,
when heat is required, produces refrigeration in the summer
time when cooling effect is to be obtained.   Other cities obtain
a certain degree of economy by utilizing the exhaust steam from
central lighting stations.   The steam is used either for heating
water used in hot-water systems or for circulating at low pressure
through systems of steam piping.

ORGANIZATION OF NATIONAL DISTRICT HEATING ASSOCIATION

As district heating has become more and more widely used,
various suggestions for economical operation have been brought
forward from time to time and the engineering fraternity has
grown to realize that in district heating lies a fruitful field for
obtaining economies through scientific construction and management.   It was in view of this situation that the need of a national
association of district heating companies became more and more
apparent, and this idea finally crystalized in an after meeting
of the Ohio Electric Light Association, July 15, 1909.   At that
time a temporary organization was formed and arrangements
were made for a convention which was held November 10, 1909,
in the Great Southern Hotel, Columbus, O.

Since that time the National District Heating Association has held conventions each year during the latter part of May or the first week in June and the association has grown from a small group of about a half-dozen men in 1909, to a membership in 1914 of about 300, including most of the heating companies in the United States, several from Canada, and also members from Germany, France, England and Russia. The objects of the association have been to promote the interests of its members in all matters relating either to steam or hot-water heating, with special reference to district heating; also to establish helpful relations with kindred associations and with manufacturers of heating equipment. The association is the outgrowth of an effort to bring the business of district heating to a practical and scientific basis and provide a general distribution of knowledge concerning the important facts connected with district heating.

That the art or business of central heating has made marked progress is shown by the steady improvement in the character of the reports and papers on steam engineering subjects provided at these annual conventions. The association each year publishes bound volumes containing these reports and their discussions by the members, and they constitute at the present time the most complete and authentic information on the subject of district heating that has thus far been published.

While like many other associations of similar character it has passed through certain periods of discouragement, it is rapidly growing to be one of the most influential technical associations in the country, and the marked spirit of loyalty and enthusiasm that pervades the association is due not only to the pleasant acquaintances and friendships formed among the members but also to the confident belief that the business of district heating has passed through its days of discouragement and that the future is laden with tremendous possibilities in the development and expansion of the industry. Already there are now between 300 and 400 heating companies in the United States representing an investment of many millions of dollars. This development, in the opinion of many of those most

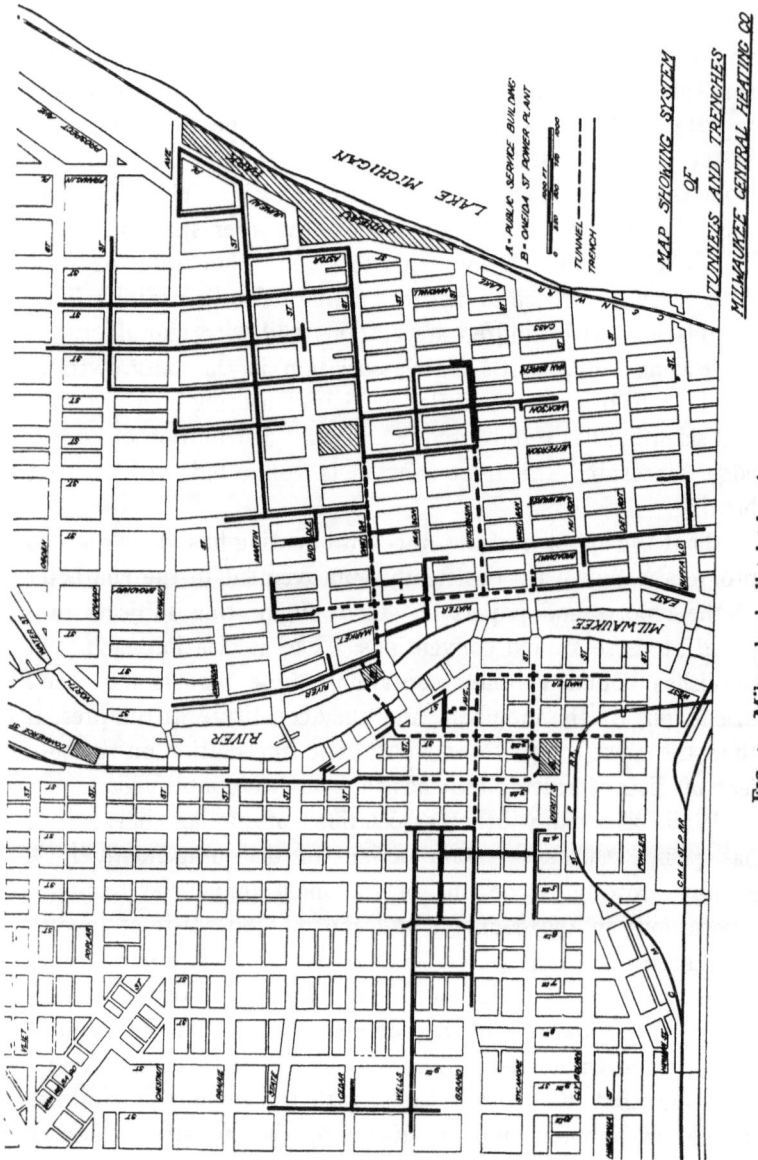

FIG. 4.—Milwaukee's district-heating system

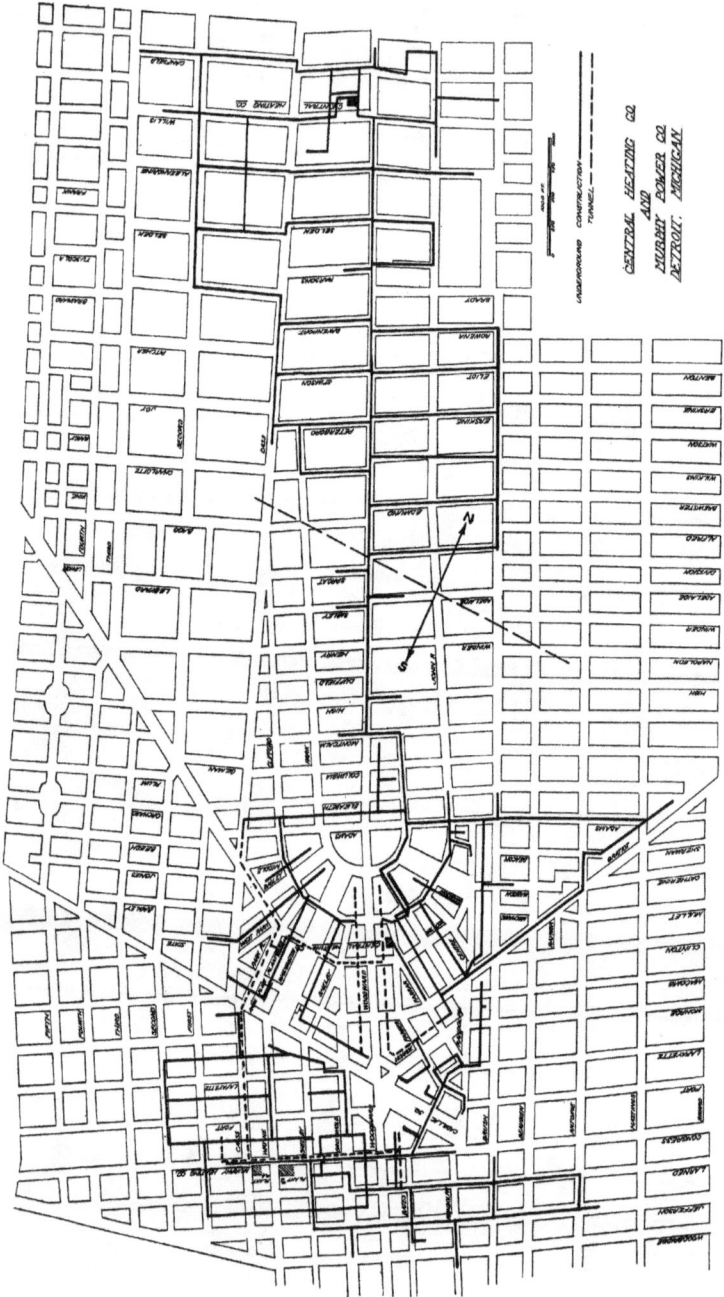

FIG. 3.—Map of the two systems of heating mains in Detroit, Mich.

CENTRAL HEATING CO.
AND
MURPHY POWER CO.
DETROIT, MICHIGAN

UNDERGROUND CONSTRUCTION
TUNNEL

To face p. 13

familiar with the subject, is merely the beginning of a great industry which will have a very important influence on future civilization not only in this country but also in other countries.

TYPICAL DISTRICT HEATING SYSTEMS

The maps given herewith, show the extent of the heating business in some of the principal cities in the country. The maps showing the heating mains in the city of New York comprise two systems: one, the downtown system in the heart of the business district; the other, the uptown system, supplying residences and moderate-sized business establishments. In September, 1914, the service supplied from these mains included 1,300 customers to whom were furnished 1,750,000,000 lbs. of steam per annum, from which the company received a yearly income of $933,000. The pipe used for these mains varies in size from 6 in. to 24 in. in diameter, and the aggregate length is about 73,000 ft.

The accompanying map showing the heating systems in Detroit, indicates two systems of mains totaling in length about 100,000 ft., supplying about 1,500 customers with a total of 1,500,000 sq.ft. of radiation. The accompanying map of the heating systems in Milwaukee shows a system of mains having a total length of about 60,000 ft., to which is already connected a half-million square feet of radiation. The accompanying map of the loop district of Chicago shows the extent of their sectional heating system. There are already about 800,000 sq.ft. of radiation connected, and if the additional steam for laundries, kitchens, elevator pumps, etc., is included, the steam load is equivalent to about 1,200,000 sq.ft. of radiation.

While the heating systems of the larger cities may be of special interest, yet the importance of district heating in the smaller cities and towns should not be overlooked. The accompanying map of the underground system of mains in the village of Pontiac, Ill., shows a steam heating system for heating residences and small stores comprising about 6,400 lin. ft. of underground pipe line. This system, which supplies 119 customers having about 80,000 sq.ft. of radiation connected is typical of a large number of small plants which, while individually somewhat limited in extent, form the separate links in an ever-increasing chain of district heating systems scattered throughout the country.

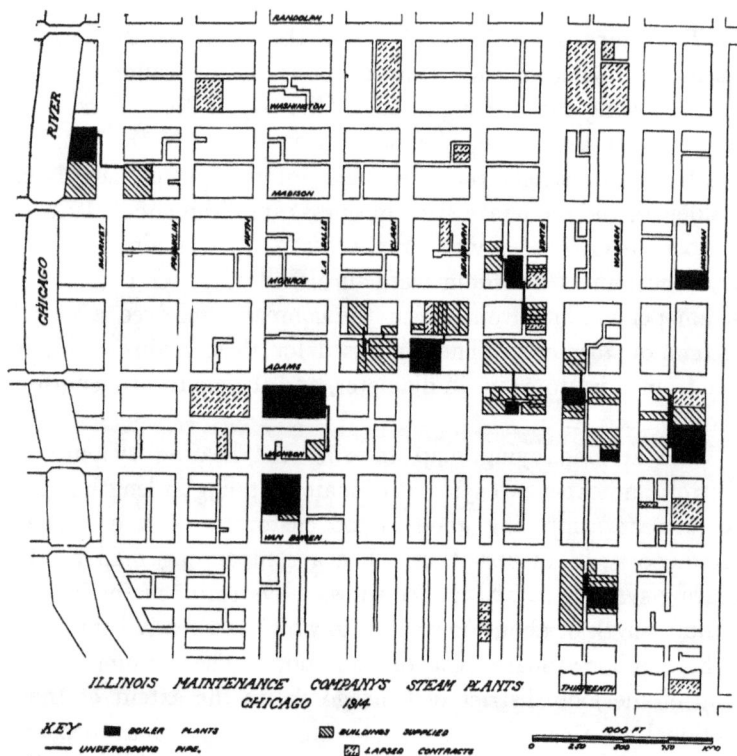

ILLINOIS MAINTENANCE COMPANYS STEAM PLANTS
CHICAGO 1914

KEY   ■ BOILER PLANTS   ⬚ BUILDINGS SUPPLIED   1000 FT
      — UNDERGROUND PIPE.   ⬚ LAPSED CONTRACTS   0   250   500   750   1000

Fig. 5 —Illinois Maintenance Company's steam plants in the loop district of Chicago

The map showing the heating mains of the Merchants' Heat & Light Co. of Indianapolis, Ind., is interesting on account of the fact that it shows not only a steam heating system in the downtown district, fed from two boiler plants on the opposite sides of the city, but also a hot-water system supplied from a single plant in the residence district. The hot-water system consists of about 52,000 lin. ft. of underground pipe line, and

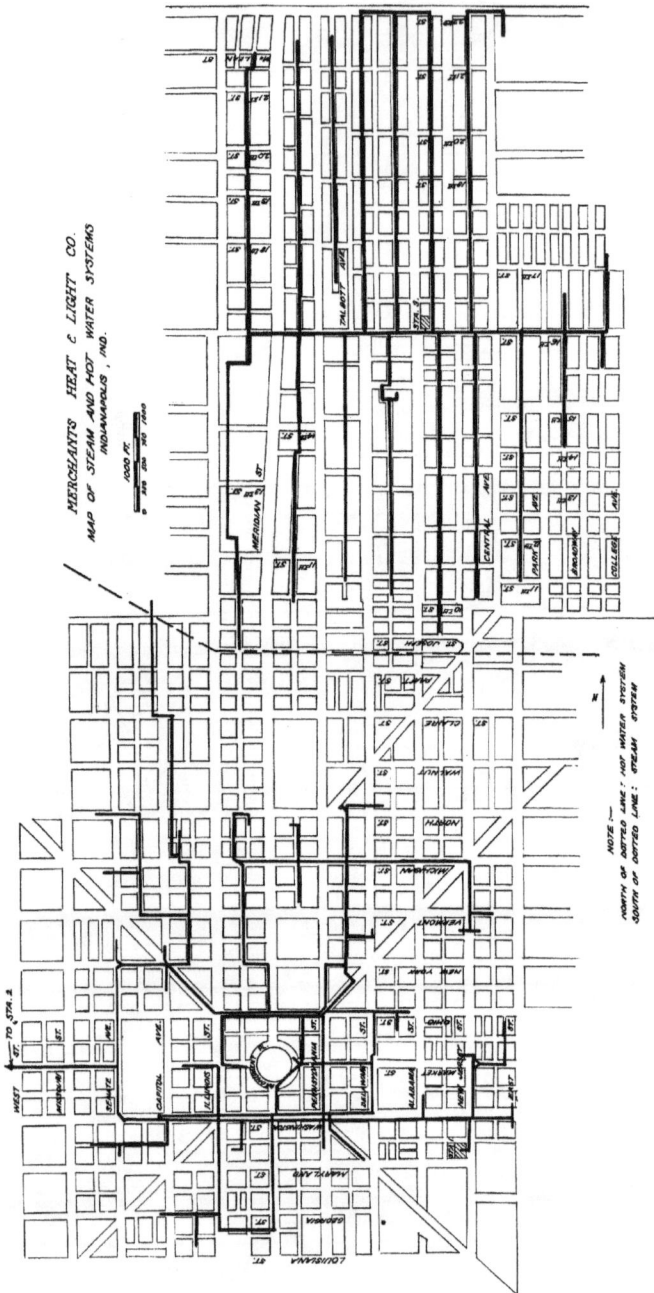

MERCHANTS HEAT & LIGHT CO.
MAP OF STEAM AND HOT WATER SYSTEMS
INDIANAPOLIS, IND.

1000 FT.

NOTE :—
NORTH OF DOTTED LINE : HOT WATER SYSTEM
SOUTH OF DOTTED LINE : STEAM SYSTEM

To face p. 15

Fig. 7.—Steam and hot-water systems of the Merchants' Heat and Light Co., Indianapolis, Ind.

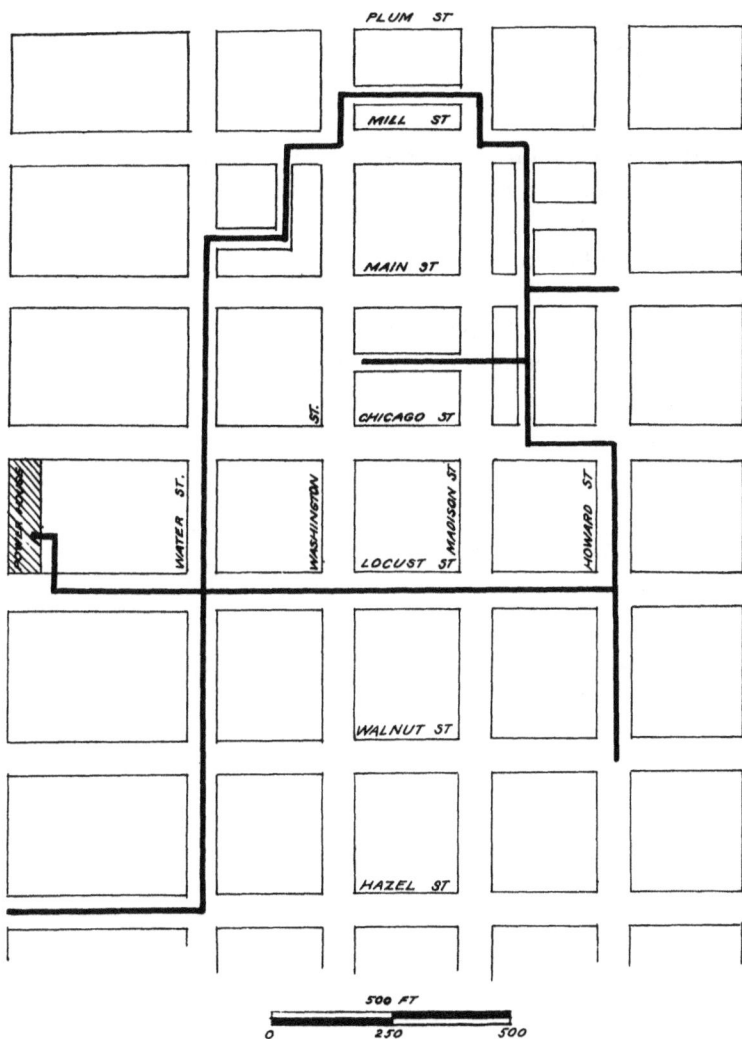

FIG. 6.—Underground system of steam mains in Pontiac, Ill.

supplies about 450,000 sq.ft. of hot-water radiation. The steam heating system consists of about 43,000 lin. ft. of underground pipe lines and supplies about 900,000 sq.ft. of radiation.

The foregoing are merely a few examples taken at random, and include only a fraction of the large and important steam heating systems already installed. The examples given, however, are probably sufficient to show that the business of district heating is rapidly becoming a large and important factor in municipal life, and is a business which has come to stay. In view of this fact, it may be interesting to take up in succeeding chapters some of the points of practical information required in the successful operation of a steam heating company.

CHAPTER II

SELLING OF HEAT

INTRODUCTION. (1) METHODS OF SELLING HEAT, AND FORMS OF CONTRACTS. (2) DISCUSSION OF METER RATES

HAVING traced the development of the district steam heating industry from its origin in 1878 up to the present time, let us now look a little more closely into the details of the business.

One of the first questions that naturally arises is, " What is the most suitable method of disposing of this commodity." A merchant sells his sugar at a unit price per pound, his cloth at so much per yard and his eggs are quoted at varying prices per dozen; but in selling heat the method of arriving at a price is not quite so simple. In the early days of district heating it was at first suggested that meters be installed to measure the exact amount of steam furnished to each building. However, the first meters were more or less inaccurate and this led to various methods of solving the problem.

Some companies would adopt a price per thousand cubic feet of space heated per season, others would base their prices on the number of square feet of radiator surface installed in the building; while still others would figure the amount of heat necessary in the building in accordance with their own formulæ and make fixed prices on this basis. All three of these methods are open to very serious objection.

It is a well-known fact that the amount of heat given off by radiators increases very rapidly when the difference in temperature between the radiation and surrounding air is increased. It can readily be seen, therefore, that in a building which is poorly supplied with radiators, much more steam would be used

per square foot of radiating surface than in a building which
is amply supplied. In other words, a building insufficiently
fsupplied with radiation will require more steam heat per square
oot of heating surface at 55° F. than a building amply supplied would need at 75° F. Moreover, if the building is amply
supplied with radiators it is necessary to shut off the steam from
many of the radiators in order to keep down the temperature.

The normal consumption of steam in radiators per season
under ordinary conditions in a climate similar to that of Chicago,
Illinois, runs from 600 to 800 lbs. of steam per square foot
where the radiating surface is properly proportioned. There
are buildings; however, which cannot be depended upon to run
very close to this average. Occasionally the consumption will
be very much above this amount in one building and in another
building it will be very much below, the amount used running
all the way from 300 to 3000 lbs. per square foot of radiating
surface during the heating season.

If a company should charge for steam service a rate of 30
cents per square foot of radiating surface per year, the price
would be equivalent to about 43 cents per thousand pounds
of steam if the consumption were 700 lbs. per square foot of
radiating surface per season.

On the other hand, the consumer who used only 300 lbs. per
square foot of radiation would pay $1.00 per thousand pounds
for his steam supply while the consumer who used 3000 lbs. per
square foot would be paying only 10 cents per thousand pounds
for his steam supply. While the average result might approximate 43 cents per thousand pounds of steam, the system of
charging is unfair to the consumer who is economical with his
steam and almost puts a premium on wastefulness.

The same situation is found where the steam is sold on the
basis of so much per thousand cubic feet of space. Two buildings may have exactly the same space enclosed within their four
walls and yet one may require double the amount of heat
necessary for the other. For example, some buildings may be
protected on two sides by adjoining buildings, so that only the
two ends of the building are acting as cooling surface, whereas

a building which is isolated will radiate heat through the walls and windows of all four sides.

In order to determine beforehand the amount of heat required by a given building, nearly every company which has gone into the steam heating business has endeavored to work out formulæ which would include the various factors which enter into the heating requirements of a building. Some of these formulæ have been used with more or less success, especially in larger buildings.

There is one factor, however, which no one has been able to embody in a formula, and that is the personal factor. One man will operate his building in such a way that by turning the steam on and off judiciously only enough heat is furnished for the necessary requirements of the occupants. Another agent by careless supervision of a building will have so much heat on at all times, that the tenants are compelled to keep their windows open with cold draughts blowing in, even in winter weather, as their only means of self defense against excessive heat.

In view of the foregoing considerations the most progressive companies are discarding the old flat rate methods of charging and are adopting systems of charging based on the actual amount of steam service used. Where heating is done by hot water, this thus far, has been impossible as meters have not been perfected for use with hot-water systems. In the case of steam service, however, meters both for measuring the steam at moderate and high pressures and also for measuring condensation have been perfected until they are very satisfactory and dependable instruments.

(1) METHODS OF SELLING HEAT AND FORMS OF CONTRACTS

From the foregoing, it will be seen that the methods of charging for steam for heating purposes may be divided into five classes.

(a) Flat rates per square foot of radiation for hot-water systems.

(b) Flat rates per square foot of radiation for steam heating systems.

(c) Flat rates per square foot of radiation, theoretically required according to the company's formulæ.

(d) Flat rate per year based on estimates of service requirements, or else based on estimate of what the customer would be willing to pay.

(e) Schedule prices based on the amount of steam used as shown by either Steam or Condensation meters.

Class "A"—Flat Rates for Hot-water Heating. Where heat is distributed by means of hot water, the price varies depending on the cost of coal in the particular locality, on the length of the heating season and on the various other factors peculiar to each individual plant. It can readily be seen that the price for heating during the heating season in the Southern States ought to be very much lower than in a State like Maine or Minnesota. The majority of heating plants are located in the Northern States and the variation in price would be approximately from 15 to 25 cents per square foot of radiation, with 20 cents as an average price in the neighborhood of Chicago, Illinois. The hot-water heating systems are exceedingly popular with the customers as they give a very mild form of heat and do not cause such sudden fluctuations in the temperature of the room, as is often found with hot-air furnaces and steam heating. Those companies which use the hot-water systems are quite strong in its endorsement, and seem to feel that it is an economical method of heat distribution.

The fact, however, that the heat cannot be sold on a meter basis according to a regular schedule as is the case with steam heating, will probably prevent a very general adoption of hot-water systems. The tendency of the times is now more and more towards exact measurements and systematic methods of dealing with central station work, and just as the old flat rate systems for lighting by incandescent lamps have passed away, meter systems are rapidly supplanting the flat rate systems in the heating business.

Where a hot-water system is used, the following may be considered a typical form of contract for the heating service:

CONTRACT FOR HOT-WATER HEATING

" This Agreement, made this ....day of..........19..
between the PUBLIC SERVICE COMPANY OF NORTHERN ILLI-
NOIS, its successors and assigns, hereinafter called the Com-
pany located in the Village of.... ... .in the State of
Illinois, and...............of the Village of..........in the
State of Illinois, hereinafter called the Consumer.

" The Company agrees to supply Hot Water for Heating
to the Consumer in the premises known as No........... ...
Street, in the Village of....... . . .upon the terms and
conditions following, to wit:

" I. The premises to be heated under this contract are
described on the accompanying diagram.

" II. The period for which the heat is to be furnished is
during the ' regular heating season ' for ....year ...,
beginning 191.., and ending .... 191...

" III. The consumer hereby agrees and guarantees that the
storm sashes and protection to the doors shall be securely
fastened in place by October 15th, and so remain until April 15th
each season. Failure by said Consumer to protect said windows
and doors during the period named shall work a forfeiture of
this contract if the Company so elect.

" IV. The Company agrees to use reasonable diligence to
maintain a comfortable temperature in said premises at all
seasons of the year during the said time, provided; that the
windows, ventilators, and doors in said building are kept properly
closed and in proper repair; that the Consumer at all times
uses reasonable diligence to the end that the best results may be
obtained from said Heating Plant; that the pipes of said Consumer are properly covered and remain so during the entire
heating season, and that the said Consumer shall place and
maintain in the premises to be heated under this contract
sufficient radiation, to be determined by the Company, to maintain the desired temperature.

" The piping up to the service cocks in the alley or street
and radiation shall be kept free from mud and scale, cared for
and maintained in good order at the expense of the Consumer
in all respects as though it were operated by said Consumer
independently, but the controlling devices except shut-off valves
in basement or on radiators, will be cared for by the Company,
any and all repairs to same to be, however, at the expense of
the Consumer. Any radiation added shall be paid for, the

same as if the radiation had been in service during the entire heating season.

" V. The water supplied shall enter the building at not less than 130 degrees, when the outside temperature is at freezing (32° F.) and shall be raised one degree for each drop of one degree outside and, provided premises are equipped with thermostat, shall be furnished in sufficient quantity to heat said building to 70° F. in the coldest weather provided the requirements of Section IV are complied with. The evidence of sufficiency of quantity shall be that the temperature of the water has not dropped more than 30 degrees below the above schedule while passing through the radiators, this to be shown by not less than four tests, taken fifteen minutes apart, in succession at the points where the water enters and leaves the premises heated.

" Complaints as to lack of sufficient heat must be made in writing stating the temperature as shown by the tests, and filed with the Company within five (5) days of its occurrence.

" VI. The Consumer agrees to pay the Company at the rate of ... . dollars per season for said heating, and payable in installments as follows: 5 percent October 1st, 15 per cent November 1st, 20 percent December 1st, 20 percent January 1st, 20 percent February 1st, 15 percent March 1st and 5 percent April 1st, in each year of the life of this contract.

" VII. The Consumer agrees to alter ... . radiation and equipment so as to make the same in accordance with the specifications of the Company, and hereby expressly releases the company from all liability for failure to properly heat the Consumer, in case said radiation and equipment are not in accordance with the specifications of the Company.

" VIII. The ' regular heating season ' is understood to be from September 15th, to June 1st, when the outside temperature is at a point below 65 degrees and heat is needed for the personal comfort of patrons of the plant.

" IX. It is understood and agreed by the parties hereto that the water in the pipes and radiators is the property of said Company, and that the drawing-off of any of said water for any purpose whatsoever shall be considered a violation of this contract and render it voidable at the option of the Company, who shall have the right to and may collect from said Consumer the full value of the water and the heat contained in it. It is further understood and agreed by the parties hereto that no range-boiler or other apparatus for the heating of water shall be connected to the system without the written consent of the

Company and that sun-parlors and sleeping-porches are not to be considered as part of the house to be heated.

"X. It is further agreed that failure on the part of either party to this contract to comply with the terms thereof for five (5) consecutive days shall work a forfeiture of this contract at the option of the other party, and in case the Consumer shall fail to pay any installments for said heating when due, the Company shall have the right at any time thereafter, upon two (2) days' notice, to disconnect and shut off the water and cease to furnish water for the heating herein provided for.

"XI. It is hereby agreed and understood that the Company shall not be liable for any damage to person or property caused by the lack of hot water sufficient to heat said premises, provided said lack of hot water be caused by accident, breakage or any other cause beyond the control of the Company, and that said Company shall not be liable for damage caused by leaks in the piping or radiation, caused by bursting pipes or otherwise.

"XII. The Company shall at all reasonable hours, by its inspectors, agents, or workmen have the right to free access to the premises supplied with heat, or to the valves or service pipes from mains, to look after the heating system, to turn the water on or off, to adjust and care for the controlling devices or other portions of the heating system of said premises, and to make any and all such tests as may be necessary or desired by the Company.

"It is understood and agreed between the parties hereto that all indorsements, schedules, or memoranda on the back of this instrument are made a part of this agreement.

"XIII. It is hereby understood and agreed that this contract is not assignable nor transferable by the Consumer without written consent of the Company and that in case of a sale or leasing of said premises by the Consumer during the life of this contract no rights under this contract shall pass to the purchaser of such sale or lease, and in case of such sale or lease the Company or its successors or assigns may at its or their option at once terminate or cancel this contract and shut off the supply of hot water to said premises.

"This contract although signed is subject to the approval of the Vice-President of the Company and shall not be binding upon the Company until indorsed with his approval.

"It is finally agreed that all the terms and stipulations heretofore made or agreed to by the parties in relation to said service are merged in this contract and that no previous or contemporaneous representations or agreements made by the Company's officers or agents shall be binding upon the Company, except as and to the extent herein contained.

PUBLIC SERVICE COMPANY OF NORTHERN ILLINOIS

By.........................
Contract Agent
....
By.........................
Approved...................19...
PUBLIC SERVICE COMPANY
OF NORTHERN ILLINOIS
By.................
Vice-President.

Class "B"—Flat Rate for Steam Heating. The next form of contract is for steam heating on the basis of the number of square feet of radiation installed. The flat rate per square foot of radiation varies according to the locality and conditions under which the plants are operated as in the case of hot-water systems. The rate, however, for steam radiation is about 50% higher than that for hot-water radiation, the price varying from about 25 to 35 cents per square foot of radiation for districts having a temperature condition approximately like that of Chicago, Illinois. The difference in rates between steam and hot-water radiation is in view of the fact that steam radiation usually transmits about 50% more heat per square foot of radiating surface than is transmitted by hot-water radiation. The heating season varies in length according to the latitude.

In districts corresponding to Chicago, Illinois, the heating season is about eight months in length, or approximately from October 1st to June 1st. The payments during the season are based on arbitrary percentages and run approximately 5% October 1st and April 1st, 15% November 1st and March 1st, 20% December 1st, January 1st and February 1st. Other similar divisions are made in different localities where steam is furnished on flat rate system. It is very important that thermostat regulators be installed in all the buildings, otherwise steam will be wasted to a considerable extent in moderate weather. It is also well to use economizers in the basement, in order to get the full benefit of the heat value in the steam. These points will be described more at length in a succeeding chapter on " distributing systems."

While there are modifications which are likely to come up in various localities, the following form of contract will cover the ground for flat rate heating for the average plant.

CONTRACT FOR STEAM HEATING SERVICE (FLAT RATE)

" AGREEMENT entered into this——— day of————, 191— between MERCHANTS' HEAT AND LIGHT Co., hereinafter called the Company and——————hereinafter called the Consumer, both of Marion County, State of Indiana.

" The Company agrees to supply the Consumer, and the Consumer agrees to take from the Company for a term of years, beginning ——— 191— and ending ———— 191-, steam for heating the building known and designated as———————

" Steam shall be furnished only during the heating season, a period of eight months, beginning September 25th, and ending May 25th, and only as may be required by the weather conditions during this period.

" The Company agrees to use reasonable diligence in providing a regular and uninterrupted supply of steam sufficient to maintain a temperature of 70 degrees within the premises at point of control, when the outside temperature is 10 degrees below zero, but does not guarantee such, and shall not be liable in damages to the Consumer for failure to temporarily supply steam to said premises, due to accidents, strikes, or other causes beyond the control of the Company, beyond a pro rata deduction for the actual time of such failure; and such claims for deductions, to be valid, shall be presented in writing at the office of the Company within forty-eight hours of the time of occurrence of such failure.

" There shall be no charge at all made by the Company for such periods when the temperature at point of control in the premises supplied, drops below 55 degrees Fahrenheit; and for such periods when the temperature may be greater than 55 degrees and less than 70 degrees a discount shall be allowed justly proportional to the loss below 70 degrees; provided that such discount shall not be allowed unless the Company shall have been notified in writing of such insufficiency of heat, and given an opportunity to discover the cause, and if due to the Company's service, to remedy the same, nor shall it be required when the cause is due to defective or insufficient radiation, or a violation of the rules and regulations of the Company for receiving and distributing the steam, or to defective construction of the building, or to any fault of the Consumer.

" The Company shall have the right to cut off the supply of steam for non-payment of bills when due, for failure to comply with its rules and regulations, and in case the Consumer allows, suffers or permits loss or waste of steam, or the condensation therefrom, or the use thereof by any other person or for any other purpose than that herein provided for; and in the event of steam being thus cut off the Consumer agrees to reimburse the Company for the entire expense of making all service connections.

" Special attachments to piping, such as hot-water heating apparatus and the like, shall be arranged for as provided in the Company's rules and regulations, and will be charged for at special prices, depending on the character of the service required.

" The charge per season for said service shall be............
.............($........) dollars, to be paid by the Consumer at the office of the Company in the following manner, to-wit; Twenty (20%) per cent of said sum shall be charged to the Consumer on the first day of October, and a like sum on the first days of November, December, January, and February, immediately following. All bills payable within ten (10) days from date.

" The rate of charge for a fractional season shall be on the following basis: Sept. 25 to Oct. 31, inclusive six (6%) per cent, of the charge for a full season; Month of November, Twelve (12%) per cent; Month of December, Twenty (20%) per cent; Month of January, Twenty (20%) per cent; Month of February, Eighteen (18%) per cent; Month of March, Thirteen (13%) per cent; Month of April, Seven (7%) per cent; May 1st to May 25th inclusive, Four (4%) per cent. No discount will be allowed for the fractional month in which the service may be commenced or discontinued.

Description of premises to be heated:

" Should any change in the exterior of the building be made during the term of this contract, such as to increase or decrease the steam condensation, a corresponding increase or decrease, respectively, in the annual charge shall be made.

" It is agreed that the Rules and Regulations of the Company, copy of which is hereto attached, shall become part of this contract.

" This contract shall be deemed to be renewed for successive periods of one year each, unless one of the parties shall, on or before thirty days prior to the expiration of the contract term, or any successive term, respectively, give written notice to the other party of his desire to have the contract terminated.

" This contract is not subject to assignment by the Consumer without the approval of the Company, but shall innure to and bind the successors and assigns of the Company.

" This contract, although it may be signed by an agent of the Company is subject to the approval of the General Manager, or other executive officer of the Company and shall not be binding on the Company until endorsed with his approval.

" It is finally agreed that all the terms and stipulations heretofore made or agreed to by the parties in relation to said heating service are merged in this contract, and that no previous or contemporaneous representations or agreements, made by the Company's officers or agents shall be binding upon the Company except as, and to the extent herein contained.

MERCHANTS' HEAT AND LIGHT COMPANY.

By....................
Agent.
....
By....................

APPROVED...............191..

General Manager, Merchants' Heat and Light Company.

RULES AND REGULATIONS FOR HEATING SERVICE

" Buildings Supplied. Steam will be supplied to buildings adjacent to the street distributing mains of the Company that are equipped in accordance with the following requirements, having sufficient radiating surface and with piping of sufficient capacity to maintain a temperature of 70 degrees Fahrenheit inside when the outside temperature is 10 degrees below zero, with a steam pressure of one pound per square inch at the service valve.

" Contracts made with Owners. Contracts for steam heating service will be made only with the owners of the buildings to be supplied, except where the system installed supplies heat to store-room only, or to premises occupied independently of, and having no connection with other sections of the same building, and through which no pipes pass supplying steam to other portions of the same building, and except also where the tenant occupies the entire building; in which case contracts will be made with the tenants.

"Service Pipe. In connection with five-year contracts, where the annual rate is $75.00 or more, the Company will provide at its own expense the service pipe from its main to a point just inside the basement of the building to be heated, if said basement extends to the curb line, otherwise to a point just inside the curb line, and place a service valve thereon; such service pipe and valve shall be and remain the property of the Company. In case the basement does not extend to the curb line the Company will extend the service pipe from the curb line into the basement at the expense of the owner or consumer.

" The Company reserves the right to supply other premises through such service pipe, should it deem it advisable, carrying the required service pipes therefor, through the basement and walls of the Consumer's building and into the basement of adjoining buildings; all damage to walls, etc. being repaired without expense to the Consumer.

" In case of building being heated by an extension of service pipe from an adjoining building, the expense of extending such service pipe shall be borne by the owner or occupant of the building for whose benefit it is made.

" In case the annual rate for heating the premises is less than $75.00 or where a contract is made for a shorter term than five years, the service pipe will be laid by the Company from its main to a point inside the basement, at the expense of the Consumer.

" Consumer to Provide and Maintain Building Equipment. The Consumer shall furnish, own and maintain all of the piping, radiators, and appliances inside the building pertaining to the heating system, and shall keep the same in good repair. The Company's inspectors and trouble-men will be at the service of the Consumer when required, for purposes of instruction, and in locating the cause of trouble and suggesting the remedy therefor, and the Consumer shall on notice from the Company promptly correct any imperfections in the system which may be discovered at any time, no matter how caused; any approval of the heating equipment which may be given either directly or by inference at the time contract is made with the Consumer, or at the time service is begun shall not relieve the Consumer of the obligation to correct any defects which may be discovered later.

"Steam Trap. The Consumer shall provide and maintain in good repair one or more standard steam traps, as may be required approved by the Company, and of sufficient capacity to properly handle the condensation from the heating system of the building, without loss of steam. Steam trap to be equipped with air valve and glass gauge, and piped with a by-pass with the necessary valves, so the trap may be temporarily disconnected from the system for making repairs as the occasion may arise.

"Should it be necessary to locate the Company's service pipe so as to drain into the basement, instead of back into the main as will usually be done, the Consumer shall also place and maintain the necessary trap for handling the condensation therein.

"Cooling Coil. There shall be connected to the discharge pipe from the steam trap a continuous cast iron cooling or economizing coil approved by the Company, the surface of which shall be approximately equal to one-fifth the radiating surface supplied with steam. This cooling coil to be connected as an indirect radiator, being enclosed with galvanized iron and supplied with cold air from the outside, or from inside the basement if approved by the Company, the heated air being conducted to some part of the building where a constant supply of heat is required, as the cooling coil can at no time be cut-off of the system. The cold air ducts leading to cooling coils shall have at least three-quarters of a square inch area, and the warm air ducts at least one square inch area, of cross section, for each square foot of cooling coil. Cold-air duct to have damper for use when steam is cut off. Register faces must have net areas equal to the ducts in which they are set. The trap and cooling coil must be so placed that the water discharged from the trap will enter the coil at the top or highest point and leave it at the bottom, the discharge pipe then being carried to a point of sufficient height to maintain the coil all the time full of water, but no higher than the trap discharge.

"The purpose of the cooling coil is to extract all of the available heat from the condensed steam, before it is discharged into the sewer. An additional charge will be made fifteen per cent. in excess of the regular rate if the cooling coil is omitted.

" Where a supply of heat is required in the basement and conditions permit, the installation of cast iron, hot water direct radiation will be permitted, and is recommended in preference to economizing coils.

" Tell-Tale. The Consumer shall provide a ' tell-tale ' for indicating the escape of steam through the trap, to consist of a suitable length of one-quarter inch pipe, connected to the water seal of cooling coil or in the event cooling coil is omitted to be connected to the discharge pipe from the trap. In case of steam blowing out through the ' tell-tale ' the Company must be immediately notified so that the cause of such waste may be remedied.

" Pipe Covering. All pipes and fittings shall be covered, and kept covered at all times, with an approved steam pipe covering not less than one inch thick, unless it is desired to use such pipes and fittings as radiating surface. As a rule, better results will be obtained by covering all the main supply pipes, and not using them as radiating surfaces. In basements and other places where the supply pipes are used as radiating surfaces, no more pipe shall be left uncovered than is necessary to properly heat the rooms through which they pass.

" Radiators. In general, radiators should be placed near doors and windows, where the cold air enters the building. Satisfactory service cannot be given unless the radiators are of the correct sizes and properly located. The Company may refuse service to premises where these conditions do not prevail.

" The radiators shall be of approved design and material, and each radiator shall have an approved automatic air-valve, which shall be maintained by the Consumer at all times in proper working order.

" Powers Regulators. As near as practicable to the Company's main service valve the Consumer shall set one or more ' Powers ' regulating valves of the proper size, as the Company may direct, for controlling the temperature in the building, the Company will supply air at an approximate pressure of fifteen pounds per square inch for operating the regulators.

" Each regulating valve shall be connected up complete with the air supply service pipe of the Company, and with the thermostat located at such point as the Company may direct. As a rule, the thermostat will be set so as to maintain a temperature of about seventy degrees but the Company does not guarantee the successful operation of the thermostat and regulating valve at all times, and shall not be held liable on this account, or for its inability at times to maintain the necessary air pressure on its system, due to leaks or breaks in the line, etc. The thermostats will be set and adjusted for the required temperature to be maintained by the Company's inspectors, and the Consumer shall not change or tamper with same in any way.

"Atmospheric systems of steam heating shall in addition to radiator traps, have a steam trap for whole system, before discharge to sewer.

"Permit for Connection. Steamfitters and others are forbidden from making connections with the Company's service pipes or valves. or in any way altering or interfering with them without a written permit from the Company which will be used only after a contract for heating has been signed by the Consumer and accepted by the Company.

"Notice of Alterations, etc. No alterations, additions to, or deductions from the piping, radiators, or any part of the heating system, and no special attachments shall be made by the Consumer without written notice to the Company and receipt from the Company of written permit specifying the particular change. or attachment to be made. All such changes and attachments shall be subject to the Company's approval after being made.

"Company's Employees to Have Access to Premises. The Company's agents or employees shall have free access to the premises of the Consumer at all reasonable hours, to inspect the heating apparatus, to ascertain whether or not it is maintained in good repair, and to inspect and repair the service pipes, valves, meters, etc., the property of the Company.

"Inspectors, agents or employees of the Company are forbidden to demand or accept any compensation for services rendered.

"No one. not an agent or employee of the Company or otherwise lawfully entitled to do so, shall be permitted by the Consumer to inspect, remove or tamper with meters, valves or appliances, registering or controlling the steam supplied to the premises.

"Doors and Windows to be Kept Closed. Reasonable care shall be exercised by the Consumer in keeping all doors, windows, ventilators. etc.. closed to prevent unnecessary waste of heat; and all exterior doors. windows, etc. shall be kept in good repair.

"Meters. In connection with all contracts on a meter basis, the Consumer shall provide a suitable location for the condensation meter. and the connections and fittings therefor, on the discharge pipe from the cooling coil or trap. Such connections and fittings shall likewise be provided by the Consumer in connection with flat rate contracts should the Company require it, that a test meter may be set at any time to check the condensation of steam in the Consumer's system.

" The Company will furnish the meter free of charge in connection with contracts where the Consumer guarantees a minimum consumption of steam sufficient to amount to seventy-five ($75.00) dollars or more per season at the contract rate; in connection with all other contracts the meter will be furnished to the Consumer by the Company at cost, and will be maintained by the Company during the term of the contract.

Class "C"—Contracts Based on Theoretical Required Radiation.—Contracts are also based on a flat rate dependent on the amount of radiation required as determined by the formulæ of the Company. In this form of contract the price is governed not only by the amount of radiation installed, which is the minimum, but also by the theoretical radiation required to heat the building, according to the formulæ adopted by the Company and which is usually filed with the Public Service Commission in those cities or states which have a commission.

There are various formulæ used in figuring radiation. One which has been adopted by a great many heating engineers and which can be used in both steam and hot-water heating is figured as follows:

Allow 1 sq.ft. of radiation to every   2 sq.ft. of glass.
Allow 1 sq.ft. of radiation to every   20 sq.ft. of exposed wall.
Allow 1 sq.ft. of radiation to every 200 cu.ft. of contents.

The sum of these amounts gives the number of square feet of radiation required for steam. For hot-water service, add 70% to the above. This formula is based on the requirements of 70° temperature inside with outside temperature 10° F. below zero.

Another rule used by some of the heating companies is to add to the glass surface 10% of the exposed wall surface and multiply this sum by 75; to this amount add the cubical con-

tents of the room. For hot-water radiation multiply this sum
by 0.011—for steam heat by 0.0055.

The formula used for hot-water service in the town of
Fowler, Indiana and filed with the Indiana State Commission
is as follows:

1 sq.ft. of radiation to every 100 cu.ft. of contents.

1 sq.ft. of radiation to every 10 sq.ft. of exposed wall.

¾ sq.ft. of radiation to each sq.ft. of opening; i.e., doors, and
windows.

Height of exposed wall is taken as distance from outside
ground line to ceiling.

Area of exposed wall is taken to include openings.

During the year 1914 the following schedule of rates and
regulations for hot-water heating in Toledo, Ohio, was filed by
the Toledo Railways and Light Company with the Ohio Utilities Commission and accepted by the Commission. The rates
are based on the following rules for determining the square feet
of radiation required to heat a given building.

"Obtain area of exposed walls in building. From this
subtract the area of window and door openings (frame measurement) in exposed walls. This remainder should be divided by
wall constant (see table) and to the result should be added the
area of window and door openings (frame measurement). This
sum should be multiplied by 75, and to this result should be
added the cubical contents of the room. This sum should be
multiplied by temperature constant (see table). The result is
the square feet of radiation (direct) required to heat building
when building is of good construction.

Temperature Constants.	Wall Constants.
0.005 —50° F.	1 for ⅞ in. wall.
0.006 —55° F.	2 for 2 in. wall.
0.007 —60° F.	3 for 4 in. wall.
0.0075—65° F.	5 for 6 to 9 in. wall.
0.0082—70° F.	7 for 9 to 12 in. wall.
0.009 —75° F.	8 for 13 to 27 in. wall.

NOTE —The temperature to be figured at the degree guaranteed by the Company in each individual contract.

" Any residence or inhabited room will have 65° F. as a minimum temperature. Any store-room or garage may have a temperature with 50° F. as the minimum.

" Any room or space having an opening which may communicate with the rooms or space to be heated, must be included in the measurement for space heated, whether radiation be installed or not in such room or space.

" The foregoing rule is for ideal conditions, requiring the minimum amount of heat for given cubical contents. To the radiation requirement thus calculated, there must be added a percentage to provide for exposed locations, bad construction, insufficient or improper repairs, and other conditions which would make the minimum radiation requirements inadequate to keep the building comfortably warmed. These conditions, which cannot be ascertained by general rule, add from five to twenty-five per cent, above the minimum or ideal requirements.

" In all cases where the conditions are not ideal, as above defined, the Company shall so notify the applicant, and shall specify the particulars in which the building is not in ideal condition, and shall indicate the percentage of radiation which must be added by reason of such condition of the building, whereupon the applicant shall have the option of making the repairs, and changes so specified, or any part of them and shall be entitled to a corresponding reduction in the per cent, of increased radiation required.

" The hot water is to be furnished in sufficient quantity to heat said building to a temperature of......° F. in coldest weather, provided that sufficient radiation be installed by the Consumer to maintain the desired temperature. Evidence to the sufficiency of quantity of water shall be that the temperature of the hot water has not dropped more than 30° while passing through the Consumer's heating system, as shown by not less than four tests taken fifteen minutes apart in succession, tests to be made at the point of entrance of service to the building. When the radiation connected to the system exceeds the amount of radiation as figured by the formula for the temperature contracted for, the drop in temperature shall exceed the 30° drop in the same percentage as the connected radiation exceeds the radiation demand as figured by the formula for the temperature contracted for.

NOTE.—The guaranteed temperature to be left blank and inserted in each individual contract as contracted for.

" The surface area of all hot-water pipes installed in basement, or other space on the premises, not included in the measurement for radiation will be charged for as radiation unless pipes be covered with an efficient pipe covering of not less than one inch thickness.

Rates—Direct Radiation

$0.20   per sq.ft. of radiation per season,   o to 500 sq.ft.
0.1888 per sq.ft. of radiation per season, 501 to 2000 sq.ft.
0.1777 per sq.ft. of radiation per season, 2001 to 5000 sq.ft.
0.1666 per sq.ft. of radiation per season, 5001 or over sq.ft.
    40% of above rates added for indirect radiation.

" Consumers connected to the hot-water service September 30th or later, will be charged 50% of the expired portion of the heating season, and at the full rate for the balance of the season.

" Where for any reason this clause becomes inoperative the following basis will be used. Figures represent per cent. of season's heating.

September 2%	December 17%	March 15%
October 5%	January 20%	April 9%
November 12%	February 16%	May 4%

" A discount of 10% allowed if paid at the office of the Company within ten days from date of bill.

Many other formulæ have been proposed at various times, reflecting the experience of various companies. These follow more or less the form:

$$R = (Gc_g + Wc_w + Vc_v) \div K$$

where

R = sq.ft. of radiation required.
G = sq.ft. of glass or door opening.
W = sq.ft. of net wall area.

V = cu.ft. of volume.

c_g, c_w, and c_v = coefficients for various differences of temperature between the outside temperature and the room.

K = Constant depending upon the rate of transmission of heat by the radiator.

The values of c_g, c_w and c_v are given in the following table which has been used for a number of years and represents a fair average of all the values available. This table is proposed by J. C. Hornung. The values for K are as follows:

280 for Low Pressure Steam.

240 for Vapor or Atmospheric Steam.

170 for Hot Water—Temp. 180° at 0° F.

TABLE OF COEFFICIENTS FOR ESTIMATING DIRECT RADIATION.

Exposure.	B.T.U. 1° Difference	Difference in Temperature Between Room and Outside. Deg. F.					
		65	70	75	80	85	90
Single Glass, Loose..	1 50	97 5	105 0	112 5	120 0	127 5	135 0
Single Glass, Medium.	1 35	87 8	94 5	101 3	108 0	114 8	121 5
Single Glass, Tight ..	1 20	78 0	84 0	90 0	96 0	102 0	108 0
Double Glass, Storm..	0 60	39 0	42 0	45 0	48 0	51 0	54 0
Vault Glass, Sidewalk .	1 50	97 5	105 0	112 5	120 0	127 5	135 0
Single Skylight.... ..	1 08	70 2	75 6	81 0	86 4	91 8	97.2
Double Skylight ...	0 60	39 0	42 0	45 0	48 0	51 0	54 0
Good Door, $\frac{1}{2}$ Glass .	0 69	44 9	48 3	51 8	55 2	58 7	62 1
Good Door, Solid. .	0 48	31 2	33 6	36 0	38 4	40 8	43 2
Poor Frame, N.-W .	0 45	29 3	31 5	33 8	36 0	38 3	40 5
Average Frame, N.-W.	0 42	27 3	29 4	31 5	33 6	35 7	37 8
Back Pl. Frame, N.-W.	0 33	21 5	23 1	24 8	26 4	28 0	29 7
8 in. Brick, N.-W . .	0 39	25 4	27 3	29 3	31 2	33 2	35 1
12 in. Brick, N.-W. .	0 30	19 5	21 0	22 5	24 0	25 5	27 0
16 in. Brick, N.-W. ..	0 24	15 6	16 8	18 0	19 2	20 4	21 6
Ceiling, Floor above..	0 30	19 5	21 0	22 5	24 0	25 5	27 0
Floor, No Floor below..	0 15	9 8	10 5	11 3	12 0	12 8	13 5
Partition, Plastered...	0 20	13 0	14 0	15 0	16 0	17 0	18 0
V-Cubical Space.							
$\frac{1}{2}$ Change, per hour	0 009	59	64	68	73	77	82
1 Change, per hour.	0 018	1 18	1 27	1 36	1 45	1 55	1 66
1$\frac{1}{2}$ Changes, per hour	0 027	1 77	1 91	2 05	2 18	2 32	2 45
2 Changes, per hour	0 036	2 36	2 55	2 73	2 91	3 09	3 27
3 Changes, per hour	0 055	3 55	3 82	4 09	4 36	4 64	4 91
4 Changes, per hour	0 073	4 73	5 09	5 45	5 82	6 18	6 55

In a paper given by Mr. A. C. Rogers, of the Toledo Railways & Light Co., at the 1913 convention of the National District Heating Association, quite a list of radiation formulæ were given and the results worked out in a typical house. The number of square feet of radiation required varied from 461 sq.ft. according to one formula, up to 727 sq. ft. in another formula, but this was probably based on 10° below zero, and the formula would be about 635 sq.ft. for a temperature of zero. In other words, these formulæ show a variation of fully $33\frac{1}{3}\%$ between the lowest and the highest. It is, therefore, very evident that where contracts are based on the theoretical required radiation that the formula to be used should be definitely specified at the time the franchise is granted, otherwise there would be no definite basis of charging. The franchise should also definitely state whether or not the house should be provided with automatic heat regulation as this would make considerable difference in the amount of heat required to be furnished by the Company. The forms of contract under Class " C " would be the same as under " A " and " B " depending on whether they are for hot water or for steam, and the total charge per year would be determined by the formula used under Class " C ". This total amount after having been determined by the formula is inserted in the contract divided so as to give certain percentages to the different months of the heating season.

Class "D"—Flat Contract. Unfortunately in a great many heating companies many of the contracts are based on a flat price, which has been arrived at in a manner similar to that of a Jew peddler in selling his wares. In other words, neither the buyer nor the seller has a very correct idea of what the service is worth, but after due discussion arrive at a price which becomes the basis of their agreement. In the early days of block heating plants, this method has been frequently used, it sometimes being left to the central station engineer and the engineer of the customer to fix up between them about what they thought was right. It is needless to say that this method of selling steam is in the main very reprehensible. Certain companies, however, have taken up the matter, especially those who are operating private steam plants in buildings, and have derived formulæ which have given very satisfactory results in figuring the heating service.   There is, however, always the temptation on the part of customers towards wastefulness where the service is based on a flat price per year and all formulæ used in figuring for such contracts should take account of this fact.   The average result in New York and Chicago is that the Consumer will use about 25% more steam when operating under flat-rate contracts, than when operating on a meter basis.   The following form of contract is one which was used formerly by the company in Chicago in making contracts for operating steam plants in buildings at a flat price.   This contract, however, has been since supplanted by a contract which uses a meter basis for charging.

"Agreement entered into this.............. day of ....191..by and between.......... .... a corporation of.................. hereinafter called the Company, of the first part and........... .... of ....hereinafter called the Consumer of the second part, Witnesseth That:

" *Whereas* the building occupied by the Consumer at........ ....in said City of...............is equipped with a steam-generating plant, also a steam-heating system throughout said building and other devices for using steam, all of which the Consumer represents to be in a complete and good and economical operating condition.

" *And Whereas* the Consumer wishes the Company to operate said steam-generating plant to furnish to the Consumer steam therefrom for certain purposes for said building and the Company has agreed to undertake the operation of same for such purposes under certain terms and conditions.

" *Now Therefore*, for and in consideration of the premises and of the mutual covenants and agreements hereinafter contained the said parties hereto agree with each other as follows:

" The Consumer agrees to turn over to the Company for the Company's use in carrying out this contract for the purposes hereinafter set forth, the said steam-generating plant together with all appurtenances and apparatus necessary to make such steam-generating plant a first class and complete steam-generating plant in actual operation.

" The Company agrees to furnish to the Consumer from the
1st day of each October to the 31st day of each May following,
during the term hereof (during ordinary business hours not
exceeding eighteen hours in any day), at the header of said
steam-generating plant the amount of steam necessary and
proper for the heating of said building for the purpose of occu-
pancy at a temperature of seventy degrees Fahrenheit, and for
heating all water necessary for ordinary lavatory and scrubbing
purposes for said building, it being understood that the steam
to be furnished by the Company as above provided shall be
sufficient for the above-mentioned purposes, only so far as the
adequacy and efficiency of the said steam-generating plant and
the steam using and conducting apparatus in Consumer's said
building will permit.

" The Company agrees that while it is operating said steam-
generating plant, it will furnish all necessary labor and fuel
for same and will handle and remove all ashes from said premises;
also that it will furnish such supplies as fire-tools, oil, and waste
for operating said steam-generating plant.

" The Company also agrees to furnish all labor for making
such petty repairs to said steam-generating plant as can be done
by the operating force on hand, with the ordinary operating tools
at hand, but all labor for making all other repairs or replacements of every kind and character shall be furnished by the
Consumer at its own expense. All material required for making
repairs, whether the labor for same is furnished by the Company
or by the Consumer hereunder, shall be furnished by the Consumer at its own expense.

" The Consumer agrees to pay the Company for all the
services to be herein performed by the Company the sum of
.... dollars, per year payable in equal installments
of..... ....dollars each on the 1st day of each month
during the term hereof.

The Company agrees that so far as practicable it will use
all reasonable endeavors to avoid violation of the ordinances of
the City of............. ..in regard to emission of smoke and
agrees to indemnify the Consumer from and against all fines
and costs levied against the Consumer by the City of..........
. ..... on account of any violation of the smoke ordinances
growing out of the negligent operation of said plant by the
Company.

" The Company shall take out, pay for and maintain boiler
insurance on said plant to the extent of Five Thousand dollars

($5000.00) on each boiler, policy to read in the name of the Company and the Consumer as their interests may appear.

" The Consumer agrees to turn over to the Company for the purpose of enabling it to exercise the rights and privileges herein contained the boiler and coal rooms in the basement of Consumer's said building and shall allow the Company all reasonable opportunity to take into said building the necessary fuel and supplies to enable it to operate said plant. The Company's employees shall also be afforded free access to said building at all times for all such purposes as may be expedient or necessary for the proper performance of the provisions and conditions of this agreement.

" The Company agrees to furnish and keep at all times while it is operating said steam-generating plant, posted in the boiler rooms, the City boiler inspector's certificate. The Company also agrees to furnish at its own expense electricity for such electric lights around and about said steam-generating plant and the boiler and coal rooms as may be necessary for the purpose of enabling it to operate the same.

''The Consumer agrees to install a separate meter to register the quantity of water used in said boilers, and the Company agrees to pay the City bills for water through such meter.

"The Company shall have the right at any time it may so desire to furnish the said supply of steam for Consumer's said building from a source or sources outside of Consumer's said building other than from said steam-generating plant in said building. The Company shall also have the right to sell steam from said plant to other parties outside of said building in the general carrying on of its steam-distributing business in that vicinity and shall be entitled to all revenues derived therefrom, but it is understood that steam so sold shall be only over and above the required amount of steam for Consumer's said building. For the purposes in this paragraph mentioned, permission is given to the Company to extend insulated pipes to and through Consumer's said building and the outer walls thereof at convenient points to be indicated by the Consumer, without expense to the Consumer.

" Inasmuch as the duty of the Company hereunder consists in generating and delivering steam to the header of said steam-generating plant, and such steam is to be used in the heating system and said other apparatus of the Consumer, such heating system and said other apparatus being entirely under the Consumer's supervision and control, it is agreed by the Consumer that it will at all times use the steam furnished by the Company in an economical and careful manner, and that the apparatus for the distribution and use of said steam shall, at all times, be kept by the Consumer in an economical and thoroughly good operating condition; and to this latter end the Consumer agrees that all risers, mains, and steam-pipes of every description, throughout the said building, except such as are designed to be used for the purpose of radiating heat, shall be properly covered and kept insulated with a suitable and sufficient non-heat conducting material, and that said entire steam-heating system and other apparatus shall be kept at all times by the Consumer in good repair. The Company shall have the right to inspect said steam-heating system and apparatus at all reasonable times.

" The Company shall not be understood as guaranteeing a constant and uninterrupted supply of steam and shall not be liable for damages which may be caused the Consumer in consequence of the Company's failure at any time to supply steam, beyond a pro rata amount for the steam the Company fails to furnish during any period of interruption.

"The Consumer hereby expressly authorizes and empowers the Company to discontinue the supply of steam hereunder whenever any bills coming due under this contract are in arrears, or upon violation by the Consumer in any substantial or material particular of any of the terms or conditions of this contract.

" The Company may at any time by arrangement with . . . . Electric Company or others, substitute electricity for the whole or part of the steam service provided for hereunder without additional expense to the Consumer.

" This agreement shall be and remain in force for the period of . . . . . . years commencing. 19   and ending 19

" *In Witness Whereof*, the parties hereto have set their hands and seals hereto the day and year first above written.

. . . .
. . . .

Class "E" Contracts Based on Meter Readings. In Class " E " the contracts are based on the amount of steam used as shown by either steam or condensation meters. This contract is used.

1st. By central steam companies who furnish a house to house service and deliver the steam from a central plant to the curb wall of the customer.

2nd. By maintenance companies which operate the boiler plants in various buildings and supply steam to the owner from his own plant on a meter basis.

The following contract illustrates the first division of this class and while the prices in the contract are too low for some localities, it will serve as an illustration of the requirements of the company selling steam on the meter basis.

" AGREEMENT entered into this......... ... ..day of ....191.., between the MILWAUKEE HEATING COMPANY, and... ...., hereinafter called the Consumer, of Milwaukee, Wis.

" (1) The Consumer requests the Company, and the Company agrees in consideration of the agreements by the Consumer hereinafter contained, to furnish to the premises .... all the steam required by the Consumer, for heating for a term of... ....

" (2) The Consumer agrees to pay the Company all bills rendered for such steam service, within ten (10) days after the date thereof, at the following rates:

First 25,000 lbs. of condensed steam used in one month, 80¢ per 1000 lbs.

Second 25,000 lbs. of condensed steam used in one month, 70¢ per 1000 lbs.

Third 25,000 lbs. of condensed steam used in one month, 60¢ per 1000 lbs.

Next 75,000 lbs. of condensed steam used in one month, 50¢ per 1000 lbs.

All over the above amounts, 45¢ per 1000 lbs.

" (3) The Company agrees to allow the Consumer a discount of 5 per cent upon all bills paid at the Company's office on or before the 10th day of the month following that in which the steam was supplied. All bills unpaid after ten (10) days shall be considered delinquent. Failure to receive monthly bill will not be considered a reason for non-payment.

" (4) The Consumer agrees to pay the Company a minimum charge to cover the expense of inspecting and maintaining meters and service connections as follows:

When steam consumed in any month is less than 1000 pounds, a minimum charge of $1.00 for the month shall be paid.

" (5) The Consumer agrees to deposit with the Company a sum that the latter may require to guarantee the payment of bills rendered under this contract or any other indebtedness to the Company.

" (6) The heating system supplied with steam under this contract, including all pipes, traps, radiators, valves, fittings, and other appliances shall be furnished by the Consumer subject to inspection and approval of the Company, and until the Company shall inspect and approve the same, it shall not be required to furnish steam to the Consumer, and the Consumer shall at all times keep such pipes, traps, radiators, valves, fittings and other appliances in proper repair and good working order. All meters and service valves as well as all service lines from street mains to lot lines, shall be furnished by and remain the property of the Company, and it is to connect all meters; but the Consumer shall connect his system to said service valve and provide a catch basin or tank into which the condensation from the meter will be discharged before entering the city sewer system, and further agrees that no repairs, alterations or additions of any kind shall be made without the written approval of the Company, after steam has been turned on.

" (7) The Company shall at any time have the right to close and seal any valve, check, stop-cock or steam passage which might in any way divert the water condensed in the Consumer's system from the meters of the Company. No such seal shall be broken without the written consent of the Company.

" (8) The Consumer agrees that the Company does not guarantee a constant supply of steam heat, and that the Company shall not be liable for damage for any failure, to supply same, or for any defect or insufficiency of any pipe, trap, valve, fitting or other appliance furnished by the Consumer.

" (9) The Consumer agrees that the Company shall have the exclusive privilege of furnishing steam heat and maintaining steam pipes for heating on the premises covered by this agreement, so long as it shall continue in force; and it is further agreed that this contract shall be deemed to be renewed, for successive periods of one (1) year each, indefinitely, unless one of the parties hereto shall, on or before thirty days prior to the expiration of any such period give written notice that this agreement shall terminate at the end of the current period.

" (10) The Consumer agrees that properly authorized agents

of the Company shall at all reasonable hours have access to the premises for the purpose of examining, repairing or removing its meters or other materials and appliances; that no one not an agent of the Company, or otherwise lawfully entitled to do so, shall be permitted to remove or tamper with same meters, materials or appliances; to provide space for and protect from injury the meters, materials and appliances on said premises, and hereby authorizes and empowers the Company to remove the meters, and all other materials and appliances installed at its expense and to cut off the supply of steam and cap the service pipe whenever the bills hereunder are delinquent, or for violation of any of the terms and conditions of this contract.

" (11) It is agreed that all terms, stipulations and agreements made by the parties in relation to said steam service are merged in this contract, and that no representations or agreements made by the Company's officers or agents shall be binding on the Company except as and to the extent contained herein. This contract shall not be binding on the Company until approved by its General Superintendent.

" (12) This agreement shall be binding upon the successors and assigns, of said Company, and upon the heirs, personal representatives, successors and assigns of said Consumer.

MILWAUKEE CENTRAL HEATING COMPANY:

In presence of

.............................(L.S.)

....
         Solicitor.

....

APPROVED

....
      General Superintendent.

" For value received, I hereby guarantee the payment of all accounts incurred by the Consumer on account of the above contract, and agree to promptly reimburse the Company for all losses sustained thereunder.

.....................(L.S.)"

....

" N.B. If a corporation is the Consumer strike out " heirs, personal representatives "; and if Consumer is an individual strike out " successors."

The following contract is a typical form for use by heating companies which operate boilers in the Consumer's own premises. This same contract can be used for furnishing steam to a customer who has no boiler by crossing out paragraphs 6 to 13 inclusive.

" AGREEMENT entered into this . ...   . . day of
.. .. 19 .between the ILLINOIS MAINTENANCE COMPANY, a corporation, hereinafter called the Company, and
.. . . hereinafter called the Consumer, both of Chicago, Illinois, WITNESSETH:

" For and in consideration of the mutual covenants and agreements hereinafter contained, the parties hereto agree with each other as follows:

" 1. The Company agrees to furnish to the Consumer and the Consumer agrees to take from the Company for and during the period of five (5) years, beginning .... .   . 19 and after such period until thirty (30) days, after receipt by either party from the other of written notice to discontinue the service, all steam, at not exceeding . ...pounds pressure per square inch, that shall be reasonably required by the Consumer during Consumer's ordinary business hours (not exceeding 18 hours in any day) for the following purposes in the premises now occupied by the Consumer at.. ....Street, Chicago, Illinois:

" For the heating of said premises at all times between October 1st of each year and June 1st of each following year, during the term hereof that artificial heat may be necessary for the comfort of the occupants of said premises.

2. The Consumer agrees to pay the Company monthly for said supply at the following rates:

RATES

" Eighty cents (80¢) per thousand pounds for the consumption up to and including 50,000 pounds in each month;
" Seventy cents (70¢) per thousand pounds for the excess consumption in such month over 50,000 pounds and up to 100,000 pounds;
" Sixty cents (60¢) per thousand pounds for the excess consumption in such month over 100,000 pounds and up to 300,000 pounds;
" Fifty cents (50¢) per thousand pounds for the excess con-

sumption in such month over 300,000 pounds and up to 600,000 pounds;

"Forty-five cents (45¢) per thousand pounds for the excess consumption in such month over 600,000 pounds and up to 1,000,000 pounds;

"Forty cents (40¢) per thousand pounds for the excess consumption is such month over 1,000,000 pounds;

all as measured by a recording steam-meter (St. John or equal), or a standard approved type of condensation meter, whichever in the judgment of the Company shall be most suitable for the requirements to be met, to be owned and installed by the Company. A pound of steam is understood to mean one pound of water evaporated into steam. The Consumer shall provide suitable space for such meter and protect same from injury, and no one not an authorized agent of the Company shall be permitted to interfere or tamper with such meter.

"3. Bills, based on monthly readings of such meter, shall be rendered by the Company monthly and shall be payable by the Consumer within ten (10) days after their dates; and the Consumer shall be entitled to a discount of ten per cent (10%) from the amount of any monthly bill paid at the Company's office on or before ten (10) days after its date.

"4. Notwithstanding anything herein contained the Consumer shall pay to the Company during the continuance of this contract a minimum monthly bill of not less than... ... ... Dollars ($ ... ...), but each minimum monthly bill shall be subject to a discount of ten per cent (10%), if paid by the Consumer at the Company's office on or before ten (10) days after its date.

"5. The Consumer agrees to take from the Company its entire requirements of steam for the above-mentioned purposes for said building during the life of this contract.

"6. The Consumer will at all times during the continuance of this contract afford the Company the free use of the steam-generating plant in Consumer's premises, with adequate boiler and coal rooms and convenient openings therefrom to the street or alley to take in the necessary fuel or supplies and to remove ashes, also with free access thereto for the Company's employees at all reasonable times; and the Company shall be at liberty at all times to use and operate said steam-generating plant in performing this contract whether in whole or in part.

"7. The Company will, while operating said steam-generating plant, furnish all labor, water, fire-tools, and operating supplies and will handle and remove all ashes from said premises, also the labor and materials for making ordinary repairs to keep said steam-generating plant in good operating condition so far as practicable, but repairs or replacements made necessary because of faulty material or workmanship in the construction of said plant, or because of explosions, fire, or other casualties occurring beyond the Company's control, shall be furnished and performed by the Consumer. All labor and materials necessary for keeping said steam-generating plant in good order and repair while the same is not being operated by the Company shall also be furnished and performed by the Consumer.

" 8. The Company will at all times operate said steam-generating plant with due and proper care to prevent, so far as practicable, its becoming deteriorated or worn out, but if said steam-generating plant is at any time during the life of this contract condemned by the public authorities or becomes so worn out as to become unfit or unsafe to operate, the same shall be replaced by the Consumer.

" 9. The Consumer will install a separate meter to register the quantity of water used in the boilers in said steam-generating plant, and the Company will pay the City bills for water consumed through such meter.

" 10. The Company will furnish and keep at all times, while it is operating said steam-generating plant, posted in said boiler room, the City Boiler Inspector's certificate and also will, while operating said steam-generating plant, keep the Consumer's boilers insured for the benefit of the Consumer and the Company as their interests may appear, in some reliable insurance company in the sum of.............. .Dollars ($ ....   ) on each boiler and shall pay all cost of such insurance.

" 11. The Company will at its own expense while operating said steam-generating plant, furnish electricity for such electric lights around and about said steam-generating plant and said boiler and coal rooms as may be necessary for the purpose of enabling it to operate the same.

" 12. While operating said steam-generating plant, the Company will, so far as practicable, comply with the ordinances of the City of Chicago in regard to the emission of smoke and will indemnify the Consumer from and against all fines and costs levied against the Consumer by the City of Chicago on account of violation of any valid smoke ordinance, except where the violation is the result of the Consumer's own act or negligence or of the Consumer's neglect or refusal, after written request having been made by the Company, to make such repairs or replacements as the Company is not hereunder obligated to make.

" 13. The Company shall have the right to furnish the supply of steam herein provided for from a source or sources outside of said premises instead of from said steam-generating plant, if the Company shall at any time so desire, and the Company may, while it is operating said steam-generating plant, sell surplus steam from same to parties outside of said premises, if the Company shall at any time so desire, and shall be entitled to all the proceeds from the sales of such steam. For the purposes in this paragraph mentioned, the Company is hereby given the right to extend steam pipes at its own expense to and through said premises and the outer walls thereof at convenient points to be indicated by the Consumer, and in a manner satisfactory to the Consumer for distributing steam to other buildings.

" 14. The Consumer will whenever requested by the Company give its frontage consent to enable the Company to lay and maintain steam pipes in the streets and alleys surrounding said premises.

" 15. The place in said premises at which the said supply of steam shall be delivered to the Consumer (hereinafter referred to as the " point of delivery "), shall be at the steam header of said steam-generating plant except when the Company is furnishing said supply from a source or sources outside of said premises, in which case the steam supply shall be delivered to the Consumer at the point where the Company's service pipes enter the outer walls of the Consumer's premises.

" 16. All steam pipes. radiators, and other steam distributing and using apparatus and machinery in Consumer's premises beyond the " point of delivery " shall be equipped and maintained and at all times kept in good order and repair by the Consumer at the Consumer's own expense, it being expressly understood that the steam to be furnished by the Company shall be sufficient for the purposes hereinabove set forth only so far as the adequacy and efficiency of the Consumer's steam using and conducting apparatus to which the Company is to furnish steam hereunder will permit.

" 17. The Company shall not be liable for any loss or injury to the Consumer resulting from the bursting of or damage to any steam piping, equipment, apparatus or machinery in said premises beyond the " point of delivery," and the Consumer shall reimburse the Company for all loss or damage suffered by the

Company, and shall indemnify the Company against liability for any injury or damage suffered by third persons resulting from such steam piping, apparatus or machinery or from any negligence on the part of the Consumer.

"18. The Company shall be entitled to have all the return waters of condensation brought back from the Consumer's said steam-heating system and other steam-using apparatus, to the "point of delivery," or if requested by the Company, to the Company's condensation meter; but if required by the Company the Consumer shall provide suitable catch basin or tank into which such condensation may be discharged and cooled before entering the city sewer system.

"19. The Company does not guarantee that the supply of steam will be at all times constant and it is agreed that any temporary cessation of the Company's service hereunder, caused by fires, strikes, casualties, breakdowns of, or injuries to machinery, or accidents shall not constitute a breach hereof on the part of the Company, and the Company shall not be liable to the Consumer for any damages resulting from any such temporary cessation of service.

"20. If the Consumer shall make default in the payment of any bill, as aforesaid, or shall violate any of the terms or conditions of this contract and after such default or violation, the Company shall deliver at said premises, addressed to the Consumer a written notice of its intention to cut-off the supply of steam on account of such default or violation, then the Company shall have the right to cut off the supply of steam and terminate this contract at the expiration of five (5) days, after giving of said notice, unless within such five (5) days the Consumer shall make good such default or violation.

"21. In case a fire shall occur in said premises, rendering them wholly unfit for the purposes of the Consumer's business, this contract shall thereupon be suspended until such time within said term of five years as the premises shall have been reconstructed and reoccupied by the Consumer for the purposes of his business.

"This contract although signed is subject to the approval of the Vice-President of the Company and shall not be binding upon the Company until endorsed with his approval.

"It is fully agreed that all the terms and stipulations heretofore made or agreed to by the parties in relation to said service of steam are merged in this contract, and that no previous or contemporaneous representations or agreements made by the

Company's officers or agents shall be binding upon the Company except as and to the extent herein contained.

ILLINOIS MAINTENANCE COMPANY
    By.............................
          ....
                                  (*Consumer*).
APPROVED....... . ...19...
                 By.........................
                          (*Official capacity*).
ILLINOIS MAINTENANCE COMPANY
    By... ....
              Vice-President.

DISCUSSION OF METER RATES

It is important to every company no matter how small, to
have a regular printed form of contract, whether the steam is
to be sold on meter basis or not.   It is more satisfactory to the
Company and more satisfactory to the Consumer.   It is always
reassuring to the Consumer to know that he is paying the regular
price which is charged to everybody and this is especially true
in the case of commodities such as steam or electricity, which
are invisible and only appear to the Consumer in the form of
heat, light or power.

Another point which should be emphasized is the advantage
already pointed out in having a metered system of charging
for steam service.   This is a point which has been at some
times open to debate.   The advocates of the flat-rate system
have argued:

" 1. That it is more satisfactory to the customer to know just
what his heating service is going to cost him, and in that case
it would be more easy to obtain business as the customer could
compare his coal bills and other expenses with the actual cost
of the heating service.

" 2. The expense of purchasing and maintaining meters and
the cost of reading them and figuring bills has been urged as
another disadvantage against the metered system."

In the early days of the steam-heating industry there was

also complaint as to the accuracy of the meters, but with the improvements that have been made in later years, it is very generally agreed that steam meters have become a fairly satisfactory instrument for measuring. It is true that considerable care must be exercised in the installation of steam and condensation meters, and the upkeep of same. The same thing, however, is also true in regard to meters for the measuring of electricity and gas, and no one now questions the advantage of metered systems for the electric and gas systems.

In reply to the above arguments it might be urged, first that while it is often satisfactory to the customer to know what his bill is going to be, he would not be so well satisfied if he knew that the central-station company was obliged to figure about a 25% increase on his bill in order to protect themselves against waste. Furthermore, while the man who is wasteful in the use of his service, might find a saving, the man who is careful and economical in the use of his service would find the prices too high and a customer would be lost, who might otherwise be obtained on the meter basis.

As to the expense of metered service the report of the meter committee to the National District Heating Association at their meeting in 1913, states, " It has been found that the cost for inspection, reading and maintenance of meters averages in the case of two typical companies, $9.08 per meter and $8.70 per meter."

Referring again to the report of the committee, there is the following statement: " We will assume on the basis of the experience of steam companies in Detroit and other places, that the average cost per customer for meters and maintenance is $10.00 per year. Let us assume that the consumption of steam per customer on a flat-rate basis is 400,000 lbs. per season. Let us assume that on this basis the average income per customer is $180.00 per year, equivalent to a rate of 45¢ per thousand pounds. Suppose the business is changed to a meter basis at a price of 60¢ per thousand pounds. If the consumption of steam shows a saving of 25%, we will have an average bill for the customer same as before of $180.00 and the net returns to

the station will be $180.00 less $10.00, or $170.00 for 300,000 lbs.
of steam, making the average net returns (deducting the expense
of metering) 56⅔¢ per thousand pounds of steam. In other
words, the average customer is paying the same amount per
year for his heating as he did before, but the average income
for the central station per unit of steam sold has increased more
than 25 per cent."

Assuming, therefore, that a company has decided to use a
metered system of charging, the next question which arises is,
" What is a fair price to charge for the service? " In order to
arrive at a solution of this question a number of different factors
must be considered. In the first place, there should be considered, what price the traffic can bear and second, what price
the company can afford to charge and still make a reasonable
profit. In this connection, it will be interesting to note the
rates that are now used by a number of typical companies.

STEAM-METER RATES IN VARIOUS CITIES

The following list of meter rates has been collected by the
American District Steam Company, and through their courtesy
is submitted herewith.

ALBA, UTAH

1st.	10,000 lbs.	Condensation per month	$0 75 per thousand pounds
next	10,000 lbs.	Condensation per month	0.70 per thousand pounds
next	10,000 lbs.	Condensation per month	0 65 per thousand pounds
next	10,000 lbs.	Condensation per month	0.60 per thousand pounds
next	60,000 lbs.	Condensation per month	0 55 per thousand pounds
All over	100,000 lbs.	Condensation per month	0 50 per thousand pounds

ATLANTA

1st.	10,000 lbs.	Condensation per month	$0 80 per thousand pounds
Next	10,000 lbs.	Condensation per month	0 70 per thousand pounds
Next	10,000 lbs.	Condensation per month	0 60 per thousand pounds
Next	20,000 lbs.	Condensation per month	0 55 per thousand pounds
Next	20,000 lbs.	Condensation per month	0 50 per thousand pounds
Next	30,000 lbs.	Condensation per month	0 45 per thousand pounds
All over	100,000 lbs.	Condensation per month	0 40 per thousand pounds

Discount 10%

BALTIMORE

2,000 to	25,000 lbs.	Condensation per month	$0.90 per thousand pounds
25,000 to	40,000 lbs.	Condensation per month	0.85 per thousand pounds
40,000 to	55,000 lbs.	Condensation per month	0.80 per thousand pounds
55,000 to	70,000 lbs.	Condensation per month	0.75 per thousand pounds
70,000 to	85,000 lbs.	Condensation per month	0.70 per thousand pounds
85,000 to	125,000 lbs.	(Non cumulative) per mo.	0.65 per thousand pounds
125,000 to	250,000 lbs.	(Non cumulative) per mo.	0.60 per thousand pounds
250,000 to	500,000 lbs.	(Non cumulative) per mo.	0.55 per thousand pounds
500,000 to	1,000,000 lbs.	(Non cumulative) per mo.	0.50 per thousand pounds
All over	1,000,000 lbs.	(Non cumulative) per mo.	0.45 per thousand pounds

Discount 5%

BIRMINGHAM

1st.	10,000 lbs.	Condensation per month	$0.85 per thousand pounds
2nd.	10,000 lbs.	Condensation per month	0.75 per thousand pounds
3rd.	10,000 lbs.	Condensation per month	0.65 per thousand pounds
Next	20,000 lbs.	Condensation per month	0.60 per thousand pounds
Next	20,000 lbs.	Condensation per month	0.55 per thousand pounds
Next	30,000 lbs.	Condensation per month	0.50 per thousand pounds
All over	100,000 lbs.	Condensation per month	0.45 per thousand pounds

10% discount on 100M or less, 15% on 101M to 250M, 20% from 251M to 500M, 25% from 501M to 750M, 30% more than 750M.

BRANDON

1st.	10,000 lbs.	Condensation per month	$1.00 per thousand pounds
Next	15,000 lbs.	Condensation per month	0.90 per thousand pounds
Next	25,000 lbs.	Condensation per month	0.85 per thousand pounds
Next	25,000 lbs.	Condensation per month	0.80 per thousand pounds
Next	25,000 lbs.	Condensation per month	0.75 per thousand pounds
Next	50,000 lbs.	Condensation per month	0.70 per thousand pounds
Next	50,000 lbs.	Condensation per month	0.65 per thousand pounds
All over	200,000 lbs.	Condensation per month	0.60 per thousand pounds

No discount.

CANTON, ILLINOIS

1,000 to	20,000 lbs.	Condensation per month	$0.75 per thousand pounds
20,000 to	40,000 lbs.	Condensation per month	0.70 per thousand pounds
40,000 to	60,000 lbs.	Condensation per month	0.65 per thousand pounds
60,000 to	100,000 lbs.	(Non cumulative) per mo.	0.60 per thousand pounds
100,000 to	150,000 lbs.	(Non cumulative) per mo.	0.55 per thousand pounds
All over	150,000 lbs.	(Non cumulative) per mo.	0.50 per thousand pounds

10% discount.

CEDAR RAPIDS

1st.	10,000 lbs.	Condensation per month	$1.00 per thousand pounds
All over	10,000 lbs.	Condensation per month	0.50 per thousand pounds

10% discount.

CHICAGO

1st.	50,000 lbs.	Condensation per month	$0.80	per thousand pounds
50,000 to	100,000 lbs.	Condensation per month	0.70	per thousand pounds
100,000 to	300,000 lbs.	Condensation per month	0.60	per thousand pounds
300,000 to	600,000 lbs.	Condensation per month	0.50	per thousand pounds
600,000 to	1,000,000 lbs.	Condensation per month	0.45	per thousand pounds
All over	1,000,000 lbs.	Condensation per month	0.40	per thousand pounds

10 % discount.

CLEVELAND

1st.	10,000 lbs.	Condensation per month	$1.00	per thousand pounds
Next	20,000 lbs.	Condensation per month	0.80	per thousand pounds
Next	20,000 lbs.	Condensation per month	0.70	per thousand pounds
Next	20,000 lbs.	Condensation per month	0.60	per thousand pounds
Next	20,000 lbs.	Condensation per month	0.50	per thousand pound
Next	200,000 lbs.	Condensation per month	0.45	per thousand pounds
All over	290,000 lbs.	Condensation per month	0.40	per thousand pounds

10% discount.

COLORADO SPRINGS

1st.	10,000 lbs.	Condensation per month	$0.70	per thousand pounds
Next	10,000 lbs.	Condensation per month	0.65	per thousand pounds
Next	10,000 lbs.	Condensation per month	0.60	per thousand pounds
Next	20,000 lbs.	Condensation per month	0.55	per thousand pounds
Next	40,000 lbs.	Condensation per month	0.50	per thousand pounds
All over	90,000 lbs.	Condensation per month	0.45	per thousand pounds

15% discount.

DAVENPORT

1st.	10,000 lbs.	Condensation per month	$0.95	per thousand pounds
Next	40,000 lbs.	Condensation per month	0.60	per thousand pounds
Next	50,000 lbs.	Condensation per month	0.55	per thousand pounds
Next	100,000 lbs.	Condensation per month	0.50	per thousand pounds
All over	200,000 lbs.	Condensation per month	0.45	per thousand pounds

10% discount.

DAYTON

1st.	10,000 lbs.	Condensation per month	$0.75	per thousand pounds
Next	10,000 lbs.	Condensation per month	0.70	per thousand pounds
Next	10,000 lbs.	Condensation per month	0.65	per thousand pounds
Next	20,000 lbs.	Condensation per month	0.60	per thousand pounds
Next	50,000 lbs.	Condensation per month	0.55	per thousand pounds
Next	300,000 lbs.	Condensation per month	0.50	per thousand pounds
All over	200,000 lbs.	Condensation per month	0.45	per thousand pounds

10% discount.

DETROIT—RESIDENCE DISTRICT

1st.	100,000 lbs.	Condensation per month	$0.60	per thousand pounds
All over	100,000 lbs.	Condensation per month	0.45	per thousand pounds

10% discount. Minimum charge $3.00.

ERIE

1st.	10,000 lbs.	Condensation per month	$0 75 per thousand pounds
Next	10,000 lbs.	Condensation per month	0 70 per thousand pounds
Next	20,000 lbs.	Condensation per month	0 65 per thousand pounds
Next	20,000 lbs.	Condensation per month	0.60 per thousand pounds
Next	30,000 lbs.	Condensation per month	0 55 per thousand pounds
Next	30,000 lbs.	Condensation per month	0.50 per thousand pounds
Next	40,000 lbs.	Condensation per month	0 45 per thousand pounds
All over	160,000 lbs.	Condensation per month	0 40 per thousand pounds

100M lbs. or less 10% disc. All over 100M lbs. 15% disc. Minimum bill of $3.00 on Buildings of 25M cu. ft. space or less.

EXCELSIOR SPRINGS, MD.

1st.	10,000 lbs.	Condensation per month	$0 90 per thousand pounds
Next	10,000 lbs.	Condensation per month	0 83 per thousand pounds
Next	10,000 lbs.	Condensation per month	0 76 per thousand pounds
Next	10,000 lbs.	Condensation per month	0 69 per thousand pounds
Next	10,000 lbs.	Condensation per month	0 62 per thousand pounds
All over	50,000 lbs.	Condensation per month	0 55 per thousand pounds

No discount. Minimum charge $5.00.

GENEVA

1st.	50,000 lbs.	Condensation per month	$0 60 per thousand pounds
Next	50,000 lbs.	Condensation per month	0 55 per thousand pounds
All over	100,000 lbs.	Condensation per month	0 50 per thousand pounds

10% discount.

INDIANAPOLIS

1st.	10,000 lbs.	Condensation per month	$0 70 per thousand pounds
2nd.	10,000 lbs.	Condensation per month	0 60 per thousand pounds
3rd to 9th.	10,000 lbs.	Condensation per month	0.50 per thousand pounds
10th to 34th.	10,000 lbs.	Condensation per month	0 45 per thousand pounds
All over	340,000 lbs.	Condensation per month	0 40 per thousand pounds

10% discount.

KANSAS CITY

1st.	50,000 lbs.	Condensation per month	$0 65 per thousand pounds
2nd.	50,000 lbs.	Condensation per month	0 55 per thousand pounds
All over	100,000 lbs.	Condensation per month	0.45 per thousand pounds

10% discount.

LAMAR

1st.	10,000 lbs.	Condensation per month	$0 80 per thousand pounds
Next	10,000 lbs.	Condensation per month	0 75 per thousand pounds
Next	10,000 lbs.	Condensation per month	0 70 per thousand pounds
Next	10,000 lbs.	Condensation per month	0 65 per thousand pounds
All over	40,000 lbs.	Condensation per month	0 60 per thousand pounds

20% discount. 5% for cash.

LARAMIE

1st.	10,000 lbs.	Condensation per month	$0 90 per thousand pounds
Next	10,000 lbs.	Condensation per month	0 80 per thousand pounds
Next	10,000 lbs.	Condensation per month	0 70 per thousand pounds
Next	100,000 lbs.	Condensation per month	0 6c per thousand pounds
Next	100,000 lbs.	Condensation per month	0 50 per thousand pounds
All over	230,000 lbs.	Condensation per month	0 40 per thousand pounds

Minimum charge $2.00.

Discounts, 10% on 150M, or less. 15% on 151M to 250M. 20% from 251M to 350M. 25% on all over 350M.

LINCOLN

1st.	20,000 lbs.	Condensation per month	$0 70 per thousand pounds
Next	20,000 lbs.	Condensation per month	0 65 per thousand pounds
Next	20,000 lbs.	Condensation per month	0 60 per thousand pounds
Next	40,000 lbs.	Condensation per month	0 55 per thousand pounds
Next	200,000 lbs.	Condensation per month	0 50 per thousand pounds
Next	200,000 lbs.	Condensation per month	0 48 per thousand pounds
All over	500,000 lbs.	Condensation per month	0 46 per thousand pounds

3% discount.

LOCKPORT

1st.	10,000 lbs.	Condensation per month	$0 70 per thousand pounds
Next	10,000 lbs.	Condensation per month	0 65 per thousand pounds
Next	10,000 lbs.	Condensation per month	0 60 per thousand pounds
Next	20,000 lbs.	Condensation per month	0 55 per thousand pounds
Next	40,000 lbs.	Condensation per month	0 50 per thousand pounds
Next	250,000 lbs.	Condensation per month	0 45 per thousand pounds
All over	340,000 lbs.	Condensation per month	0 40 per thousand pounds

15% discount.

MARYVILLE

1st.	10,000 lbs.	Condensation per month	$0 70 per thousand pounds
Next	10,000 lbs.	Condensation per month	0 65 per thousand pounds
Next	10,000 lbs.	Condensation per month	0 60 per thousand pounds
Next	20,000 lbs.	Condensation per month	0 55 per thousand pounds
Next	40,000 lbs.	Condensation per month	0 50 per thousand pounds
Next	250,000 lbs.	Condensation per month	0 45 per thousand pounds
All over	340,000 lbs.	Condensation per month	0 40 per thousand pounds

10% discount. Minimum charge $5.00.

MILWAUKEE

1st.	25,000 lbs.	Condensation per month	$0 80 per thousand pounds
Next	25,000 lbs.	Condensation per month	0 70 per thousand pounds
Next	25,000 lbs	Condensation per month	0 60 per thousand pounds
Next	75,000 lbs.	Condensation per month	0 50 per thousand pounds
All over	150,000 lbs.	Condensation per month	0 45 per thousand pounds

5% discount.

MISSOULA

1st.	10,000 lbs.	Condensation per month	$1 00 per thousand pounds
Next	25,000 lbs.	Condensation per month	0 90 per thousand pounds
Next	100,000 lbs.	Condensation per month	0 80 per thousand pounds
All over	135,000 lbs.	Condensation per month	0 70 per thousand pounds

No discount. Minimum charge $.15 per M. cu. ft. space.

MONTREAL

Under	25,000 lbs.	Condensation per month	$0 75 per thousand pounds
25,000 to	50,000 lbs.	Condensation per month	0 70 per thousand pounds
50,000 to	100,000 lbs.	Condensation per month	0 65 per thousand pounds
100,000 to	200,000 lbs.	Condensation per month	0 60 per thousand pounds
200,000 to	300,000 lbs.	(Non-cumulative) per mo.	0 55 per thousand pounds
All over	300,000 lbs.	(Non cumulative) per mo.	0 50 per thousand pounds

No discount.

NEWBURGH

1st.	10,000 lbs.	Condensation per month	$0 60 per thousand pounds
Next	10,000 lbs.	Condensation per month	0 55 per thousand pounds
Next	10,000 lbs.	Condensation per month	0 50 per thousand pounds
Next	20,000 lbs.	Condensation per month	0 45 per thousand pounds
Next	40,000 lbs.	Condensation per month	0 425 per thousand pounds
Next	250,000 lbs.	Condensation per month	0 375 per thousand pounds
All over	340,000 lbs.	Condensation per month	0 35 per thousand pounds

5% discount.

O'NEIL

1st.	10,000 lbs.	Condensation per month	0 85 per thousand pounds
Next	30,000 lbs.	Condensation per month	0 80 per thousand pounds
Next	60,000 lbs.	Condensation per month	0 75 per thousand pounds
Next	200,000 lbs.	Condensation per month	0 70 per thousand pounds
All over	300,000 lbs.	Condensation per month	0 65 per thousand pounds

No discount.

OSKALOOSA

1st.	5,000 lbs.	Condensation per month	$0 95 per thousand pounds
Next	5,000 lbs.	Condensation per month	0 75 per thousand pounds
Next	10,000 lbs.	Condensation per month	0 65 per thousand pounds
Next	10,000 lbs.	Condensation per month	0 60 per thousand pounds
Next	20,000 lbs.	Condensation per month	0 55 per thousand pounds
Next	50,000 lbs.	Condensation per month	0 50 per thousand pounds
Next	250,000 lbs.	Condensation per month	0 45 per thousand pounds

No discount.

Buildings containing 25,000 cu ft. space or less, monthly minimum charge $3 00 net   Larger buildings 12c net per 1000, cu ft. space per month.   Churches, theatres, etc., or buildings heated occasionally, 6c net per 1,000 cu. ft. space per month.

OTTUMWA

1st.	5,000 lbs.	Condensation per month	$0 95 per thousand pounds
2nd.	5,000 lbs.	Condensation per month	0 75 per thousand pounds
2nd.	10,000 lbs.	Condensation per month	0 65 per thousand pounds
3rd.	10,000 lbs.	Condensation per month	0 60 per thousand pounds
4th &5th.	10,000 lbs.	Condensation per month	0 55 per thousand pounds
6,7,8&9th.	10,000 lbs.	Condensation per month	0 50 per thousand pounds
Next	250,000 lbs.	Condensation per month	0 45 per thousand pounds

No discount.

PEORIA

1st.	4,000 lbs.	Condensation per month	$1.50 per thousand pounds
Next	100,000 lbs.	Condensation per month	0.50 per thousand pounds
Next	200,000 lbs.	Condensation per month	0.45 per thousand pounds
All over	304,000 lbs.	Condensation per month	0 40 per thousand pounds

10% discount.

PENN YAN

1st.	50,000 lbs.	Condensation per month	$0.60 per thousand pounds
Next	50,000 lbs.	Condensation per month	0 55 per thousand pounds
All over	100,000 lbs.	Condensation per month	0.50 per thousand pounds

10% discount.

PORTLAND

1st.	5,000 lbs.	Condensation per month	$1.00 per thousand pounds
Next	10,000 lbs.	Condensation per month	0 95 per thousand pounds
Next	10,000 lbs.	Condensation per month	0 85 per thousand pounds
Next	25,000 lbs.	Condensation per month	0 75 per thousand pounds
Next	50,000 lbs.	Condensation per month	0.65 per thousand pounds
Next	100,000 lbs.	Condensation per month	0.55 per thousand pounds
Next	300,000 lbs.	Condensation per month	0.50 per thousand pounds
All over	500,000 lbs.	Condensation per month	0.45 per thousand pounds

10% discount.   Minimum charge $5.00.

SEATTLE

1st.	50,000 lbs.	Condensation per month	$0 75 per thousand pounds
Next	50,000 lbs.	Condensation per month	0 65 per thousand pounds
Next	100,000 lbs.	Condensation per month	0.50 per thousand pounds
Next	300,000 lbs.	Condensation per month	0 45 per thousand pounds
Next	500,000 lbs.	Condensation per month	0 43 per thousand pounds
For more than 1,000 M.		Condensation per month	0.43 per thousand pounds

No discount.   Minimum charge $3.85.

SEDALIA

1st.	10,000 lbs.	Condensation per month	$0 90 per thousand pounds
Next	10,000 lbs.	Condensation per month	0 85 per thousand pounds
Next	10,000 lbs.	Condensation per month	0 80 per thousand pounds
Next	20,000 lbs.	Condensation per month	0 75 per thousand pounds
Next	20,000 lbs.	Condensation per month	0 70 per thousand pounds
Next	20,000 lbs.	Condensation per month	0 65 per thousand pounds
Next	250,000 lbs.	Condensation per month	0 60 per thousand pounds
Next	260,000 lbs.	Condensation per month	0 55 per thousand pounds
All over	600,000 lbs.	Condensation per month	0 50 per thousand pounds

Discount 10%.
Monthly bills amounting to $3.35 or more—10% discount.
Buildings containing 25,000 cu. ft. space or less, monthly minimum charge $3.00 net. Larger buildings 12c net per 1,000 cu. ft. space per month. Churches, Theatres, etc., heated occasionally, 6c net per 1,000 cu. ft. space per month.

SIOUX CITY

1st.	10,000 lbs.	Condensation per month	$0 90 per thousand pounds
Next	10,000 lbs.	Condensation per month	0 80 per thousand pounds
Next	10,000 lbs.	Condensation per month	0 70 per thousand pounds
Next	10,000 lbs.	Condensation per month	0 60 per thousand pounds
All over	40,000 lbs.	Condensation per month	0 50 per thousand pounds

5% discount.

SPRINGFIELD, MO.

1st.	10,000 lbs.	Condensation per month	$0 60 per thousand pounds
Next	10,000 lbs.	Condensation per month	0 50 per thousand pounds
Next	70,000 lbs.	Condensation per month	0 45 per thousand pounds
Next	250,000 lbs.	Condensation per month	0 40 per thousand pounds
All over	340,000 lbs.	Condensation per month	0.35 per thousand pounds

10% discount.

ST. JOSEPH

	5,000 lbs. or less Condensation per month	$0.80 per thousand pounds
Over	5,000 lbs. and under 10,001 lbs. Cond.	0.70 per thousand pounds
Over	10,000 lbs. and under 30,001 lbs. Cond.	0.65 per thousand pounds
Over	30,000 lbs. and under 100,001 lbs. Cond.	0.55 per thousand pounds
Over	100,000 lbs. and under 200,001 lbs. Cond.	0 50 per thousand pounds
Over	200,000 lbs. and under 250,001 lbs. Cond.	0 45 per thousand pounds
Over	250,000 lbs. and under 500,001 lbs. (non-cum.)	0.40 per thousand pounds
All over	500,000 lbs. (non-cum.)	0 35 per thousand pounds

Discount 10%. Minimum charge $2.00 per month. 1912.

ST. PAUL

Under	30,000 lbs.	Condensation per month	$0 90 per thousand pounds
30,000 to	60,000 lbs.	Condensation per month	0 85 per thousand pounds
60,000 to	90,000 lbs.	Condensation per month	0.80 per thousand pounds
90,000 to	120,000 lbs.	Condensation per month	0 75 per thousand pounds
120,000 to	150,000 lbs.	(Non cumulative) per mo.	0.70 per thousand pounds
All over	150,000 lbs.	(Non cumulative) per mo.	0 60 per thousand pounds

10% discount.

In looking over the above list of rates, taking as the average consumption 50,000 lbs. per month, it will be interesting to see what would be the average of all the above rates for a consumer of this size. By tabulating all the above rates and taking the average, the average net price for 50,000 lbs. is found to be $.6319 per thousand pounds. If this represents the average central station bill, it would mean that the average income of the steam system would be about on this basis. From the standpoint of what the business would bear, it is evident that this rate is low, as a large number of stations have demonstrated their ability to obtain a higher rate. This is also shown from the fact that many stations find that they have no difficulty whatever in reaching their capacity in steam-heating business and that the difficulty is in supplying the business which comes to them, rather than in obtaining additional business. This fact shows also that with proper sales organization a much higher rate could probably be obtained with a satisfactory degree of patronage; also on the basis of the cost of production, the rate appears low.

The very able report given by Mr. A. D. Spencer, before the National District Heating Association in 1910, showed that the total cost of steam production in both the A. and B. plants of the Detroit Edison Company was approximately 60¢ per thousands pounds of steam delivered. Mr. Byron T. Gifford on page 81 of his recent work on " Central Station Heating," figures out a very interesting example showing the cost of steam production in an average plant of 200,000 sq.ft. of radiation and selling 120,000,000 lbs. of steam per year. He figures that in order to take care of the operating costs and fixed charges it would be necessary to have an income of about $74,344.00 per year for 120,000,000 lbs. of steam which would be equivalent to a rate of about 62¢ per thousand pounds. Mr. Gifford's example, however, is based on a plant which is operating practically on full load, but the average central heating-plant cannot expect to grow very rapidly without installing heating mains and boiler plant apparatus to take care of future loads. This means that a reasonable amount should be added to the rate to take care of the period between the installation of the plant and full load conditions.

For example, the Central Heating Company of Milwaukee, with very large and extensive systems of underground mains has thus far connected only 500,000 sq.ft. of radiation or only a fraction of the future load on the system. If we examine the rate of the Milwaukee Company, we find that the average rate for the 50,000 lb. customer would be about $71\frac{1}{4}$¢ per thousand pounds. The fact that this company has already built up a large business on this rate shows that it is not too high a rate to secure business and at the same time, there is a reasonable surplus between their rate and the theoretical cost of steam production to allow for losses during the building up of the business.

The New York Steam Company has already been referred to as the largest steam company in the United States and their rate for 50,000 lbs. per month is even higher, nearly 90¢ per thousand pounds.

It might be interesting at this point to note the schedule of prices for steam of the New York Steam Company. According to their schedule the steam is sold at fixed rates per thousand Kals (a Kal being one pound of water evaporated into steam).

The rates per thousand Kals are based on the average weekly consumption during the month in accordance with the " Graduated Scale of Prices," of the company.

The following is an extract from the scale of prices for a 4 weeks bill:

MINIMUM CHARGE $10.00

5,000 Kals	$2.00	per thousand	$10.00
6,400 Kals	1.75	per thousand	11.20
8,400 Kals	1.50	per thousand	12.60
13,400 Kals	1.37	per thousand	18.36
18,400 Kals	1.24	per thousand	22.82
23,200 Kals	1.142	per thousand	26.49
39,600 Kals	0.961	per thousand	38.06
53,600 Kals	0.867	per thousand	46.47

72,000 Kals	0.786 per thousand	56.59
99,600 Kals	0.702 per thousand	69.92
138,200 Kals	0.639 per thousand	88.31
179,400 Kals	0.60 per thousand	107.64
208,000 Kals	0.58 per thousand	120.64
254,000 Kals	0.555 per thousand	140.97
309,600 Kals	0.533 per thousand	165.02
400,000 Kals	0.51 per thousand	204.00
500,000 Kals	0.50 per thousand	250.00

In addition to the regular meter charge, a charge is made for a portion of the condensation in service and connecting pipes leading from the street mains when such pipes are of unusual length. Neglecting this extra charge, the accompanying Fig. 8 shows a curve plotted from the New York rates, together with curves showing the rates in Chicago and Milwaukee. It may be noticed that the Chicago rate curve is much more gradual than the others. This is due to the fact that the Chicago company sometimes operates comparatively small heating boilers in buildings and it is therefore necessary to keep up the rates for moderate steam consumptions. The Milwaukee schedule offers a fair compromise between the New York and Chicago rates and is perhaps a good example of a judicious system of rates for the average town. However, the cost of coal, labor and other elements in steam production should be carefully studied in each individual locality before attempting to prescribe the rate schedule.

A mistake that has been made by some companies in arranging a schedule of rates has been to suppose that it is necessary to make the cost of steam correspond with the cost of coal to the small consumers. While the Consumer is likely to make some objection to any increase in his expenses, he soon finds that the convenience and availability of outside steam supply makes it much more desirable than heat from a small boiler. There is no doubt that it is much cheaper for the ordinary householder to use the old-fashioned kerosene lamps than the modern methods of gas and electricity.

Few, however, will be found willing to go back to the old methods of lighting. The tendency of the times is towards the reduction of the various items of work around the house and just as the housekeeper is glad to be rid of the attention required by candles and kerosene lamps so a short experience with district

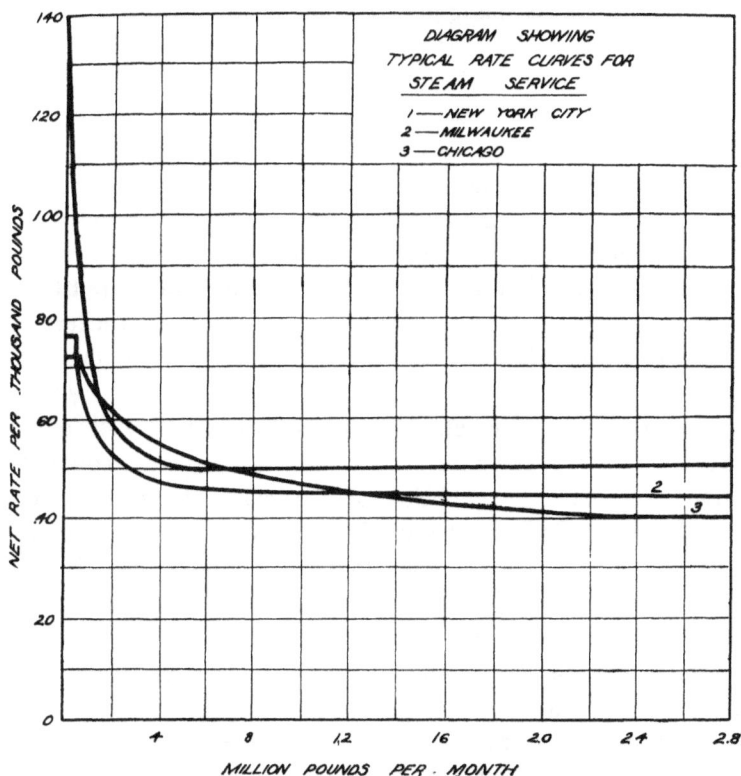

FIG. 8.—Diagram showing typical rate curves for steam service in New York, Milwaukee and Chicago

heating makes one very reluctant to return to the old scheme of independent boilers or furnaces requiring constant attention on the part of the people in the house. Things which formerly seemed luxuries are rapidly growing to be considered necessities.

Therefore, in comparing the cost of central heating service with the use of a small independent boiler, it will be necessary at all times to include the value of the labor expended in caring for the boiler, not to mention the disadvantage of dirt and soot which always accompany the delivery of coal and the shoveling of same into a furnace. The fact that not only small residences, but a great many large business blocks are adopting central-station heating shows that the advantages of this method of heat supply are becoming more and more apparent.

The fact that the heat is so convenient and that it can be obtained by simply turning on a valve sometimes leads to waste-fulness, and it is often necessary for the central heating company to caution their customers and show them ways in which they can economize in the use of heat. Where the heat is purchased on a meter basis, these suggestions are invariably received by the consumer in the most friendly spirit, as they not only show him how to keep his bills within a satisfactory range but they also make him feel that the company is trying to look after his interests rather than to sell the largest possible amount of steam.

Before leaving the subject of rates, it might be well to call attention to the two-rate system, which has been frequently advocated, viz.: the adoption of a primary charge based on the theoretical amount of radiation connected and a secondary charge based on the meter readings. This kind of rate is per-haps more thoroughly sound than the single sliding-schedule, due to the fact that the primary rate can be made to closely approximate the investment charge, while the secondary rate can be based on the operating costs. The chief objection to this rate is that it requires that the theoretical radiation for each customer be figured from a basic formula and such esti-mates of theoretical radiation required are more open to con-troversy than the reading of a satisfactory meter. Another point in favor of the simple sliding-schedule based on meter readings is that it is more easy to explain this method of charging to the customer.

While it is very possible that the two-rate system will be the future basis for the sale of steam, just as it is already to a large extent the basis for the sale of electricity, yet it is questionable as to whether the time for this change has arrived.

CHAPTER III

HEAT DISTRIBUTING SYSTEMS

(a) STREET SYSTEMS. (b) METHODS OF PIPING WITHIN BUILDINGS

In taking up the subject of Heat Distribution by means of steam or hot water, the subject naturally divides itself into two divisions, consisting first of the piping required in the streets and second of the piping required in the buildings supplied.

(a) STREET SYSTEMS

The disposal of any commodity requires efficient methods of delivery to the consumer with the minimum loss in transit. The principal loss of economy in district heating lies in the necessity for transmitting heat—steam or hot water—through pipes and conduits under the streets. Even in the best designed systems this function is attended with considerable waste, and to the neglect of this fact may be attributed a large number of the failures which have occurred in this industry in the earlier days of experimentation. Due to carelessness and inadequate design, there are many companies which have been unable to pay proper dividends, whereas a proper understanding of the essential points would have reversed the situation. A careful survey of the great majority of plants will show that it is possible to attain financial success if the proper precautions are observed, and if a suitable system of rates is combined with a properly arranged system of pipe lines.

There are several essential points to consider in the design of heat distribution systems, and these will be taken up in the following order:

1. Location of station.
2. Territory served.
3. Location of mains.
4. Topography and soil.
5. Type of system.
6. Pressures and temperatures.
7. Sizes of pipe lines.
8. Type of conduit.
9. Form of contract for pipe-line construction.

1. Location of Station. The conditions determining the location of central heating plants are usually the same as those that govern any other generating station, with the exception that it is unnecessary to build near a source of water for condensing operation. An effort should be made to select a location as near as possible to the center of the heating system, since by so doing the losses of distribution may be reduced. However, this is obviously affected by the values of property and by disadvantages in placing a plant where the territory is congested. The most logical course would be to locate near a railroad, keeping as close to the center of the system as practicable. In case the system can utilize the exhaust steam from an electric power-generating plant, the location is predetermined and need not enter into the calculations unless the plant is so far removed from the territory to be served that distribution losses will be prohibitive.

2. Territory Served. It will be assumed that there has been some demand created by the public for the installation of a central plant and that the promoters have estimated the approximate number of buildings which stand ready and willing to use the new form of heating. When the engineer is called in his preliminary survey will be expected to include an estimate of the probable growth of business within the immediate and near future. His estimate should show the number of buildings, already heated by means of steam or hot water radiation. Experience has shown that from 70% to 90% of the householders in first-class residence districts will become users of the service in due time, that is to say, in from 3 to 5 years, or as soon as the value of the service is demonstrated beyond a doubt. The greatest source of error lies in extending heating mains into poor territory, in hope of future improvement; the small margin of profit in the business may be seriously impaired or wiped out entirely by injudicious extension of service lines, as it is necessary to secure income from every foot of underground construction to assure success of the venture. The extension of the system should proceed slowly and systematically. Of course, vacant property in the better class neighborhoods will in the nature of things be expected to improve and companies are sometimes justified in providing facilities for taking care of the future loads in such cases.

In estimating the probable connected load in business districts it should be remembered that here the station will meet more or less competition from private plants in large buildings, and therefore the percentage of this class of buildings to be counted on should be decreased to perhaps 50% or less. In very large cities this percentage will be even lower as a general rule for the reasons already given. Detailed information of the amount of radiation installed should be entered upon a plat of the district to be served, this being necessary for an intelligent design of the distributing system.

3. Location of Mains. The method of installing the distributing mains and lateral or service connections should be studied. The most prevalent custom is to use the streets for this purpose, notwithstanding the generally increased cost of construction and maintenance, due primarily to the removal of paving. The company is usually required by its ordinances or contracts to extend the service pipe to the curb line or lot line, and if the street is very wide, it will be seen that the alley presents the least costly method, especially if the alley is not paved. However, it has been found in many cities that construction work in alleys is hampered by the relatively cramped space as compared with the streets. Street construction always means more or less interference with traffic, which is a serious item in large cities, but of less consequence in smaller towns.

Another consideration favoring alley construction is in the avoidance of numerous other public utility supplies,—gas, water, electricity, telephone, telegraph, sewer and street railway conduits. Unless a municipality maintains a well-organized department for regulating the installation of these utilities, or requires tunnels or galleries for same, the ground will be found honeycombed with a maze of pipes and conduits, which render impossible any consideration of a very comprehensive system of distribution, unless the rates received for service are sufficient to cover a comparatively large amount for fixed charges.

It has been the practice in many cases, especially where high-pressure steam is distributed to utilize the basement space or sidewalk space to connect adjoining buildings, the only underground work necessary consisting in crossing streets and alleys with short connecting lines. With this arrangement, it is permissible to use an expensive form of tunnel construction in order to have every part of the line available for inspection and repair in case of an accident of any nature. This is the method advised for block-heating plants in large cities. It is obviously unsafe to rely on this latter method entirely as it may involve complicated legal arrangements between the company and the various owners of the premises through which the lines are run, especially in the case of long term contracts. In residence districts the pipe may be laid under the park-way or lawn space between the sidewalk and curb.

4. Topography and Soil. The character of the soil and its topography have an important bearing on the design of a heat-distributing system. With level territory either a hot-water system or a steam system is permissible. If the general contour shows a rise in elevation away from the power-house so that the most remote consumer is well above this elevation, then if desired the returns from a steam system may be very profitably returned to the plant, the only drawback to this being in the relatively shorter life of return lines, usually necessitating their replacement long before the steam-pipe has deteriorated to that extent. The practicability of hot-water systems where the plant is located below the general level of the consumer's premises depends upon the allowable pressures which may be economically carried under such conditions. Increasing pressures introduce several unsatisfactory elements into the design, which would be much aggravated if the city is located in a hilly territory. To summarize therefore, it will be seen that:

Flat and level territory— is adapted to.
  { Hot-water systems.
  Single pipe steam systems.
  Two-pipe steam systems with
    vacuum pumps to return
    the condensation.

General sloping ground with plant at low level—is adapted to.
  { Hot-water systems, if difference in elevation does not necessitate excessive pressures.
  One or two pipe steam.

Hilly or rolling territory—is adapted to.   One-pipe steam systems only.

As before remarked the nature of soil has an important bearing on the design and on the selection of the type of conduit. One of the greatest causes for deterioration in underground construction is seepage of water into the pipe-line casings. This can be prevented only by adopting suitable precautions as to construction and materials and by installing a first-class system of drainage. It is customary to run a drain-tile line under the conduit for receiving the seepage and transmitting same to proper sewer connections, unless the soil in that locality consists of sand and gravel and the natural drainage is away from the locality.

5. Type of System. District heating companies use two distinct methods of distribution, viz.: hot water and steam. The arrangement of these two methods may be expressed in tabulated form as follows:

Type	Service	Source of Heat.
1. Hot Water { One-pipe circuit, loop or belt system.	{ Services on shunt circuit.	(a) Live steam in closed heaters.
		(b) Circulating water heated by exhaust from engines in closed heater or surface condenser.
Two-pipe supply and return, or multiple system.	{ Services in multiple	(c) Direct water heating boilers or economizers.
		(d) Steam injectors or mixers in open heater.
		(e) Returns circulated through a jet condenser or commingler

2.
Steam
{
Single-pipe system. { Returns wasted to sewer.

Two-pipe return system. { Gravity return / Vacuum return
}

{
(a) Live steam from boilers.
(b) Exhaust steam from engines.
(c) Combination of (a) and (b).
}

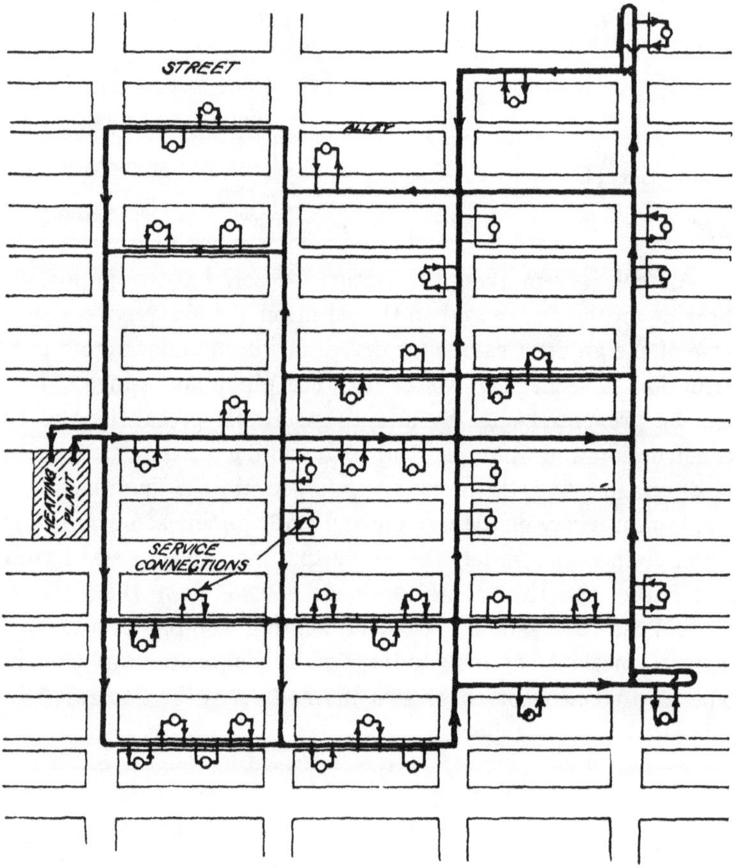

FIG. 9.—Typical one-pipe hot-water distribution system

Figs. 9, 10 and 11 show, respectively, the one and two pipe hot-water systems and a one-pipe steam system. Both steam and hot-water systems have certain advantages, the one over the other. The choice of a system depends largely on the relative weight or importance given to the different points. Some of the more important considerations relative to the different systems are included in the list given below:

Hot Water.	Steam.	Advantage.
(1) Two small pipes or one large pipe.	One large pipe for exhaust steam or smaller pipe for high pressure and return lines.	Equal.
(2) Pumping equipment should be installed in duplicate.	None required.	Steam.
(3) Service cannot be metered.	Service may and should always be metered.	Steam.
(4) High pressures and considerable leakage in pipes.	Lower pressures and little leakage.	Steam.
(5) High pressures on radiators.	Lower pressures on radiators.	Steam.
(6) Heaters required in addition to steam boilers.	No heaters.	Steam.
(7) Return water saved and brought back to heaters.	Returns are wasted due to the difficulty of maintaining return pipes.	Hot water.
(8) Radiation loss is low and expansion and contraction reduced.	Radiation loss is high due to high temperatures.	Hot water.
(9) Thermostatic regulation is necessary because meters cannot be used.	Thermostats are not necessary unless customer desires for his own economy.	Steam.
(10) Regulation is much easier.	Regulation difficult without thermostats.	Hot water.
(11) Pressure becomes too great for high buildings unless "booster" pumps are installed.	Buildings of any height may be heated on very low pressures.	Steam.
(12) Higher pressures are an element of danger.	Small liability of damage.	Steam.
(13) Cannot bead apted to steam radiation.	Can be used for any type of radiator.	Steam.
(14) Cannot be used except for heating purposes.	Can be used for other purposes than heating—cooking, laundry, pumping and industrial purposes.	Steam.
(15) Not advisable for indirect radiation because of liability of freezing.	Suitable for indirect or blast coils.	Steam.
(16) Cost of auxiliaries high.	Cost of auxiliaries lower.	Steam.
(17) Absence of noise.	High velocities create noises and water hammer.	Hot water.
(18) Hot water heat is more popular among consumers because of its low even temperature.	High temperatures with steam causing dry air and the odor of burned dust and escaping steam.	Hot water.
(19) Requires about 50% more radiation than steam.	Cost of radiation lower.	Steam.

Hot Water.	Steam.	Advantage.
(20) Does not require economy coils.	Economizers are required by many companies for cooling the returns.	Hot water.
(21) No sewer connection.	Trouble with hot return water in sewers.	Hot water.
(22) No steam traps or space for meter.	Steam traps required.	Hot water.
(23) Plant can operate condensing most of the time.	Can operate condensing only during the non-heating season.	Hot water.
(24) No cost for meters.	Cost of meters and maintenance.	Hot water.
(25) Customer does not control his service	Customer may control his service.	Steam.

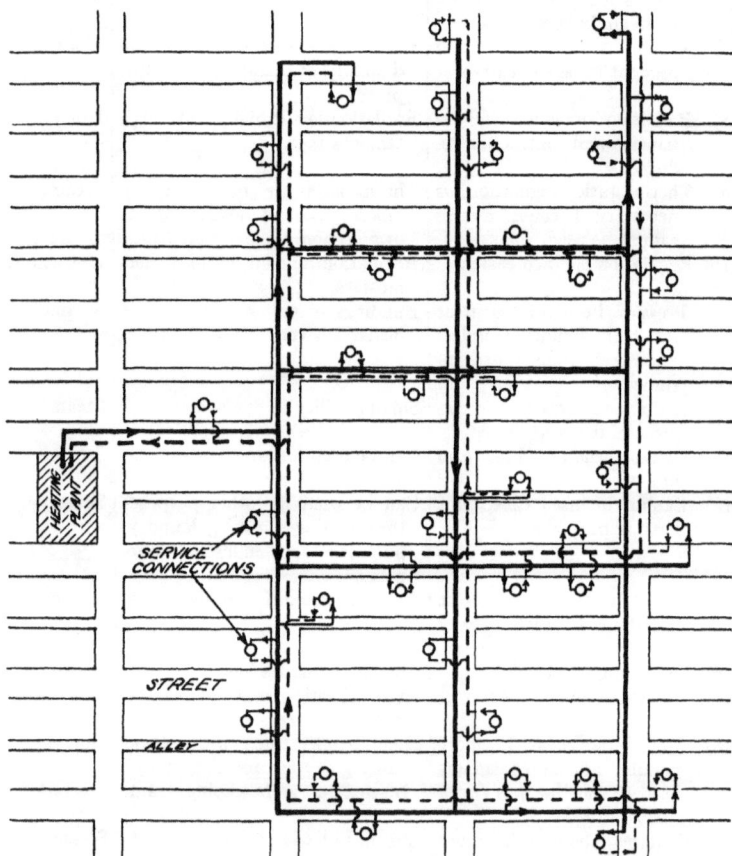

FIG. 10.—Typical two-pipe hot-water distribution system

After a careful survey of the above considerations, a decision will be arrived at for a proposed system, whether it be hot water or steam. The former system has always found favor with small residence consumers, because of its many desirable features. This preference might extend to larger buildings, but it has, until

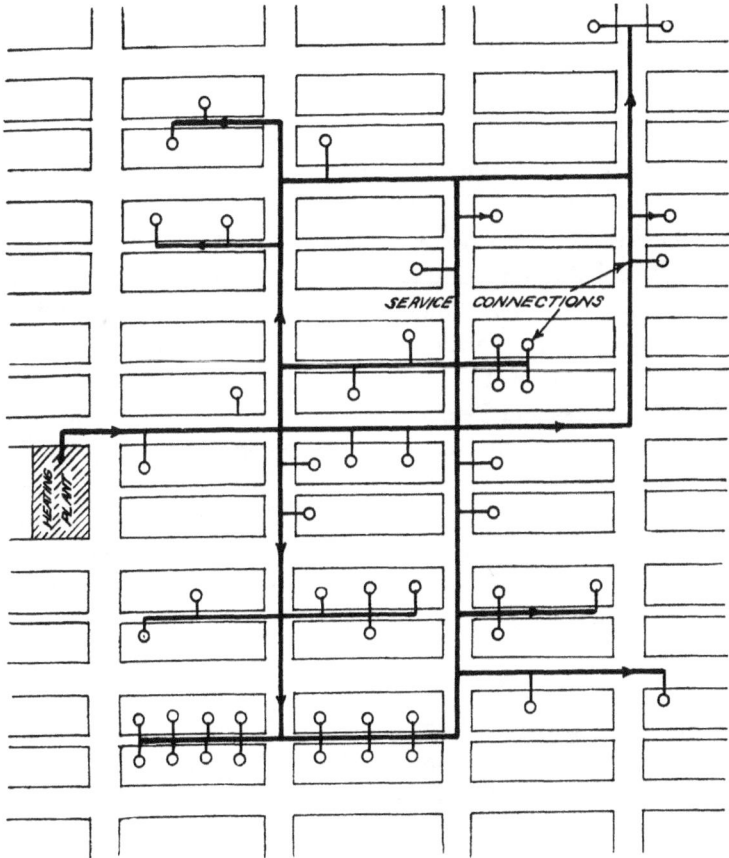

Fig 11.—Typical one-pipe steam-distribution system

the last few years, been the exception, to find a building of any considerable height equipped for hot water heat. There are many instances of late, however, where large industrial plants have installed this system. Notable among these is a new plant recently erected in Chicago, where 200,000 sq.ft. of radiating surface—a medium sized central heating plant in itself—is supplied by a two-pipe system having 9 in. supply and return lines, under a circulating pressure of 50 lbs. and a differential of 30 lbs. per square inch. In view of its size, this plant may be considered a central or district-heating system. The distributing lines are run in commodious underground tunnels, and the plant has been designed for an ultimate connected load of over 400,000 sq.ft. of radiation. Fig. 12 shows the general scheme of the piping and the method of operation will be more fully described in Chapter V, Steam-Generating Stations.

The first installations of district hot-water systems followed only a few years after steam heating was introduced, but by far the greater number of plants now installed are steam systems. This is probably largely due to the fact that the steam service can be metered. The economies inherent in the metered system of charging have accordingly swung the scale towards preponderance of steam systems.

6. Pressures and Temperatures. The question of steam pressures in steam-distributing systems is determined largely by the kind of service which is required. In residence districts where steam is needed for heating only a pressure of from 3 to 5 lbs. is sufficient. In business districts where steam is required for cooking, and technical purposes, it is often necessary to furnish a pressure of from 30 to 40 lbs. while in some districts it is necessary to bring the pressure up to 80 or 90 lbs., in order to supply steam power for elevators, pumps, etc. Where the exhaust from electric power stations is used for heating, the pressure to be carried is of course limited by the back pressure to be allowed on the engines or turbines. Where conduit construction is used in business districts the pressure allowable for satisfactory operation should not exceed 30 or 40 lbs. Where tunnel construction is used or mains for transmitting steam are run in basements or under sidewalks, higher pressures ranging from 80 to 100 lbs. can be used successfully, if lines are properly designed and suitable care is taken in turning steam off and on the system.

It is customary with hot-water systems to limit the static pressure to 100 ft. of head or 44 lbs. per square inch. Higher

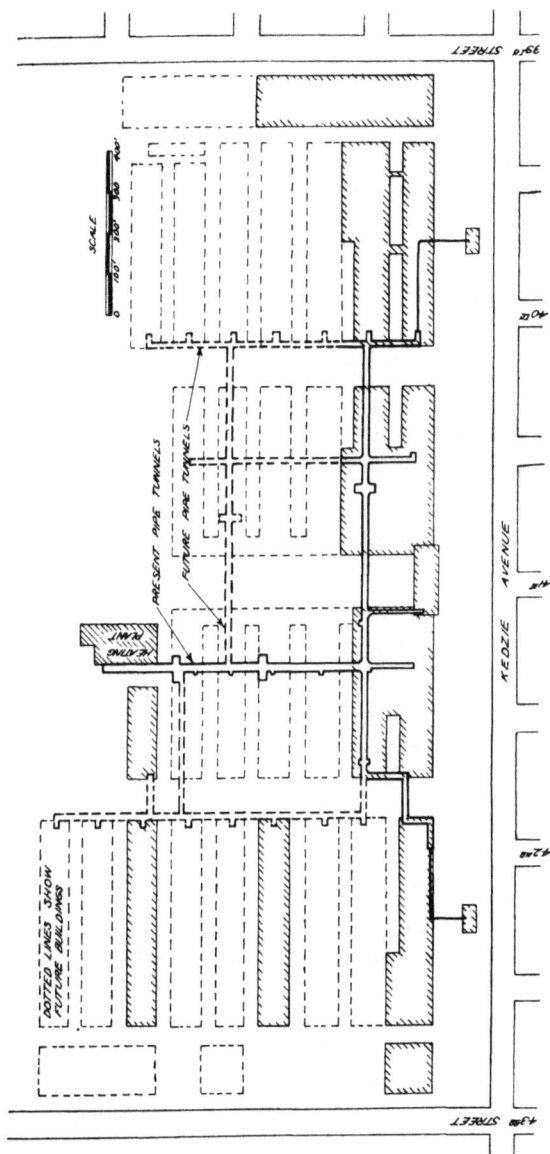

FIG. 12.—Hot-water heating system for large industrial plant

pressure increases the danger from leakage and danger from rupture of radiators. The pump discharge ranges from 40 to 70 lbs. and the return pressure from 5 to 20 lbs. per square inch. When the system is operating at high capacity, the differential pressure is greatest, the drop in temperature in the main supply line is very slight, and those services nearest the pumping plant obtain the higher outflowing pressures and lowest inflowing or return pressures, except in the case of " first flow last return " systems.

The differential pressure should never be less than $1\frac{1}{2}$ lb. per square inch at the end of the line; i.e. at the last service outlet on any lateral. The lateral service for the average residence consumer is a $1\frac{1}{4}$ in. pipe. Those nearest the plant must be fitted with choke-disks in order to reduce the inflow pressure, and prevent an excess of water flowing into the radiation. Those at the end of the line do not require these devices.

When operating the plant during zero temperature it is customary to pump the water at a speed which will provide about 6 lbs. of water per square foot of radiation per hour. The temperature of the water leaving the plant ranges from 180° to 190° F. Temperature of water returning to plant ranges from 150° to 160°, making a temperature differential of from 25° to 30° between the water entering the premises of a given consumer and that going out. When it is desired to increase the heating effect of the radiation, the greatest effect is secured by increasing the temperature of the water where practicable, rather than by increasing the speed of circulation. Hoffman's "Handbook for Heating and Ventilating Engineers " gives the following values of fair operating conditions as to pressure:

Head in feet.... 60
Static pressures, pounds per square inch......... 25
Outgoing pressure at pump, lbs. per square inch.. 50
Return pressure at pump, lbs. per square inch.... 5
Differential pressure at pump, lbs. per square inch 45

7. Sizes of Pipe Lines.   It has been the experience with many heating companies that the main trunk lines from the plant have been overloaded long before it has been expected, therefore in order to take care of the future demand it is sometimes necessary to anticipate the probable growth of the business for some years to come. While this of course operates to reduce future investment in auxiliary feeders, it involves other less desirable features in the nature of operating losses. If the company installs mains of too large diameter and the expected business is not obtained, it is easily seen that the percentage of line losses due to radiation from the conduit will be a relatively larger ratio to the total steam supplied than would be the case where the pipe is carrying nearer its capacity.

For this reason it is difficult to determine with any degree of accuracy the so-called efficiency of any particular type of construction, until the exact amount of steam passing through has been determined, and its relation established with some acceptable standard of capacity. For example, the loss from condensation due to radiation may reach 100% if no steam is being used by the various customers. This is the condition where it is obligatory for the company to carry full steam pressure on its lines regardless of the temperature, so that the steam may be " on tap " for any individual on the lines. Manifestly the actual flow is approximately zero, but for a given pressure there is a certain rate of condensation under this condition.

Now, if the line is working at full capacity the percentage of loss may be reduced to a comparatively small amount. The per cent of line loss varies inversely as the flow of steam. To reduce this loss to a minimum would seemingly be best accomplished by installing pipes of small circumference and area in order that they may be worked at high capacity at all times.

There are, however, other factors which must enter into the calculation of this problem. One of these is friction. An increase. in velocity results in a higher friction loss, which causes a drop in pressure and the liberation of some heat in passing from one pressure to another. This heat energy is not an entire loss from the pipe since it does useful work in re-evaporating moisture in the steam and bringing it again to a relatively superheated condition; and similarly in hot-water lines, in reducing the temperature drop.

However, the drop in pressure should not be excessive since this implies a higher initial pressure with consequent increased temperature and radiation loss, excessive back pressure on the engines in by-product systems, and also an increased cost of circulation in hot-water systems. A mean must be arrived at where the allowable friction loss and pressure drop is considered in relation to the radiation loss and some basis arrived at for designing the trunk lines in the system. However, in general, the radiation loss is so large in proportion to the friction loss that it is advisable to use the smallest pipe size consistent with the available initial pressure and allowable drop. The friction loss is much greater for water than for steam lines.

A fair knowledge of the natural laws pertaining to the flow of fluids in conduits is necessary for the production of an intelligently designed distribution system. Many piping systems have been failures because of the haphazard methods used and a lack of anything approximating a scientific treatment of the problem. Many of these failures could have been prevented had a sufficient amount of reliable data been at hand to foretell the performance and capacity of the lines. The absence of such precautions results in a system which is insufficient in capacity from the outset or one in which excess allowances have been made on the score of safety, involving a waste not only in initial investment but also in operating losses. The science of engineering has now reached such a stage that the relations between theoretical and practical phenomena have been formulated with considerable accuracy and therefore the errors made in old plants need not be repeated.

As this subject has been taken up at considerable length in standard text-books treating of the flow of fluids, such as water and steam, in pipe-lines and the necessary allowances to be made for bends, valves and other obstructions, it will not be necessary to go into this matter in detail. In Kent's " Mechanical Engineers' Pocket-book " will be found formulæ treating on this subject. Other books in which the subject has been discussed at length are works on heating and ventilating by Professors Hoffman, Carpenter and others. An excellent work recently published on this subject is the " Mechanics of Heating and Ventilating " by Konrad Meier, which gives charts furnishing many valuable data as to the proper design of both hot water and steam central-supply systems.

8. Type of Conduit. Figs. 13 to 15 illustrate several forms of conduit used by various companies. Generally speaking, the pipe itself will usually outlast the surrounding conduit, although the very latest forms of construction bid fair to change this condition and promise more durable qualities. The subject of insulation for underground mains is of exceptional importance. A wise selection from among the many methods now in the market must be made if efficiency in the distributing system is to be obtained. No matter how cheaply steam may be produced or how much effort is expended in power plant economies, it is of no avail if the valuable product is permitted to fade away through defective conduit before reaching the consumer. A leak in the distributing main is just as costly as it would be in any other part of the plant.

The following points should be weighed in comparing the merits of different coverings:

Character of soil, dry, wet or shifting.
Drainage conditions.
Depth below level of pavement.
Interference with other utilities.
Allowable temperatures and pressures.

In the first installation of underground steam-piping, the covering consisted of a bored log having an internal diameter a few inches larger than the pipe. This log was then coated with coal-tar on the outer surface to prevent decay. A tile-drain was laid under the casing to carry away the seepage to sewers. The sections were secured together by male and female joints. Instances are on record of these conduits being in service more than twenty-five years, the only deterioration being in the charring of the inside of the log for an inch or so.

The wood conduits shown in Figs. 13 (a), (b) and (c) were used in the early days of the industry. They are open to serious

objection and are now seldom used. Segmental wood casings
and laminated sectional wood casing, see Fig. 13 (*d*) and (*e*), are

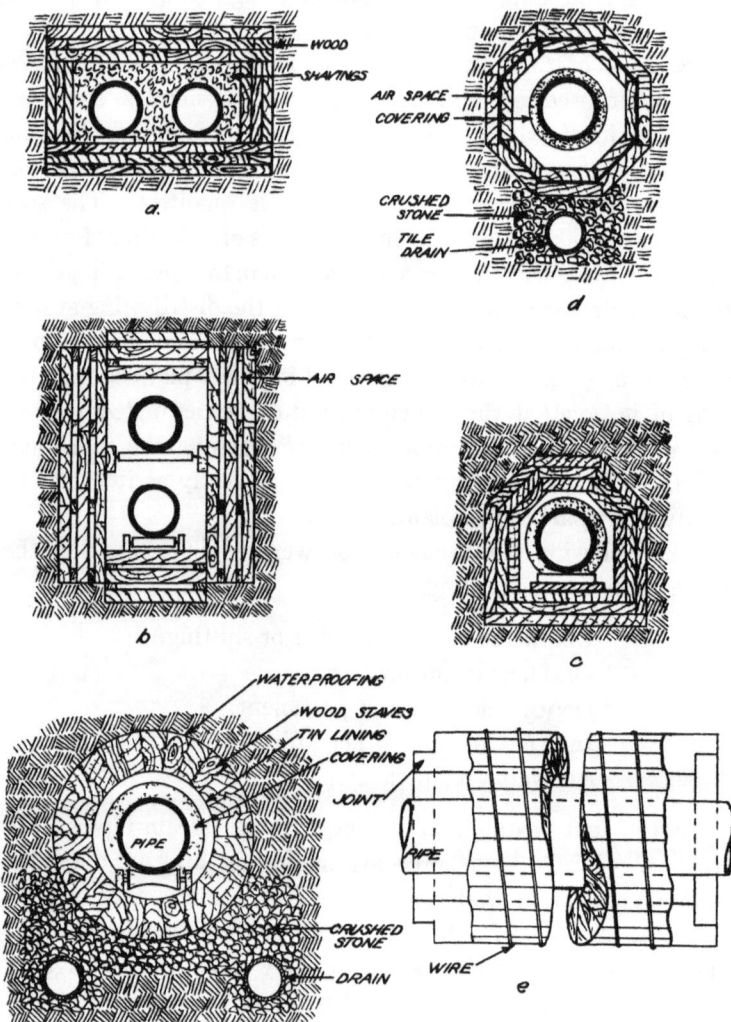

TYPES OF WOOD CONDUIT

Fig. 13.—Types of wood conduit

very popular with many engineers. These are built up from
staves of selected and seasoned white pine or tamarack and in

the smaller sizes the primitive method of bored log is still adhered
to.   The thickness of casing varies from 2 in. to 4 in., and for
very large sizes up to 6 in.   The staves are bound together by a

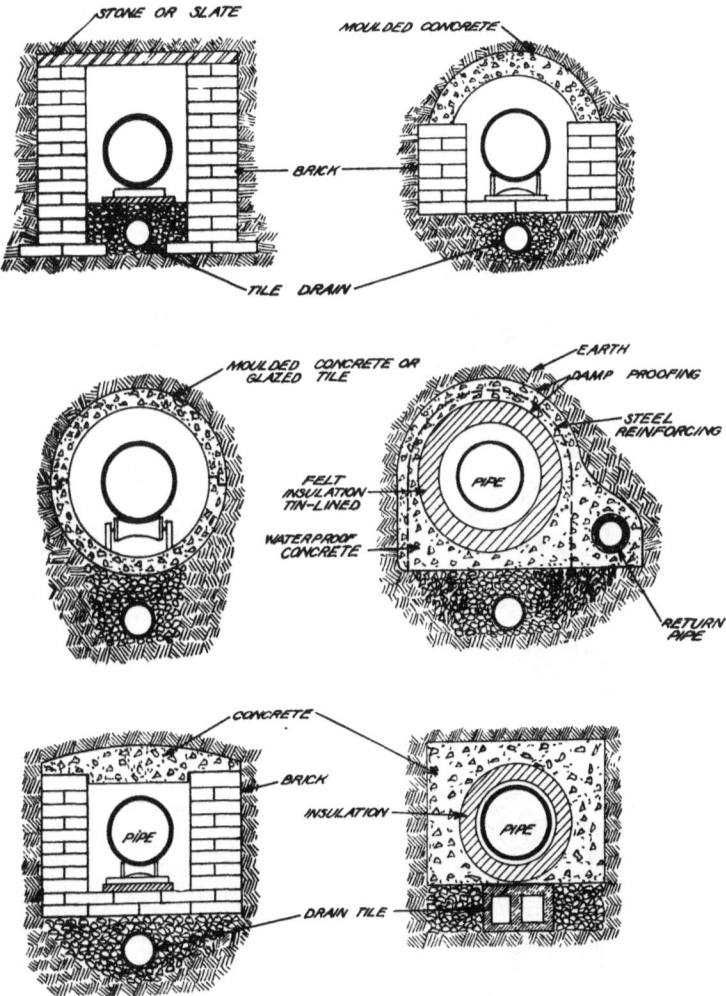

FIG. 14.—Types of brick and concrete conduit

continuous wire band.   The interior is lined with bright tin
which reflects the heat back upon the pipe.   The outside is
made water-proof by rolling in asphaltum or tar preparation.

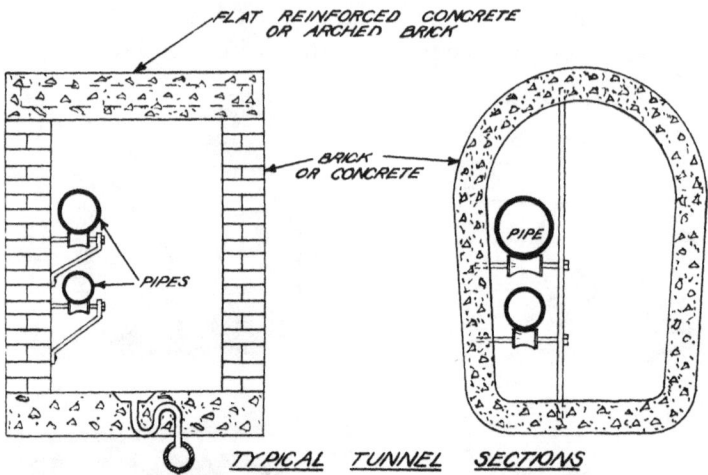

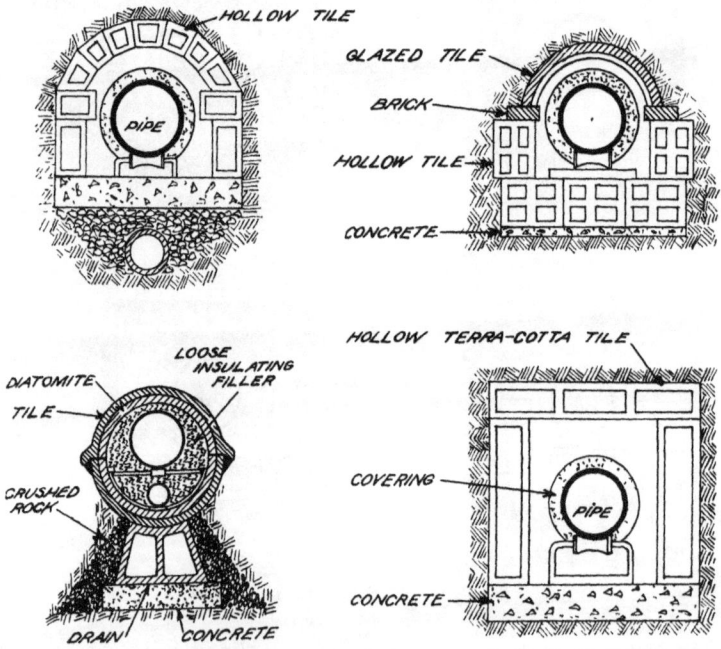

FIG. 15 —Types of tunnel and tile conduit

This makes a very efficient and durable conduit and is very largely employed at the present day. Wood pipe is also used as piping for water service, and vacuum return lines, and when used for these purposes, the friction is said to be less than in metal pipes. Wood pipe is easily installed and is easily removed in case of trouble with pipe-lines.

Many failures were recorded against wood conduit due largely to improper installations and this led engineers to devise other means of insulation for underground work. The conduits were in some cases built of brick and this type was followed by concrete and tile forms, of which several illustrations are given, Figs. 14 and 15. All conduits should have suitable under-drainage to take care of the seepage and surface water, and if moisture is kept from the conduit, its life is greatly prolonged and in many cases may outlast the life of the pipe which it protects.

For high-pressure lines, it is advisable to run the pipes in tunnels, at least the main distributing feeders, since this method offers the only guaranty against frequent shut downs due to leaks and accidents. Tunnel construction, of course, is usually out of the question for any extended system because of the high cost of installation. This is the main reason why high-pressure heating in large cities is seldom attempted on the same comprehensive basis as heating in small towns where often a single station supplies the entire community.

It is unsafe to give any data pertaining to the cost of underground construction, since this depends so much upon local conditions. In the report of the Underground Conduit Committee of the National District Heating Association, 1914, much pertinent information is to be found and the reader is advised to consult same in case detailed figures are desired.

The latest and best forms of construction may be figured at comparatively low rates of depreciation, 3 to 5% depending upon the mode of investing the sinking fund and without allowing for changes and enlargements which are very likely to occur in growing systems.

9. Form of Contract for Steam Pipe-Lines. Many companies engaged in the business of installing district-heating systems

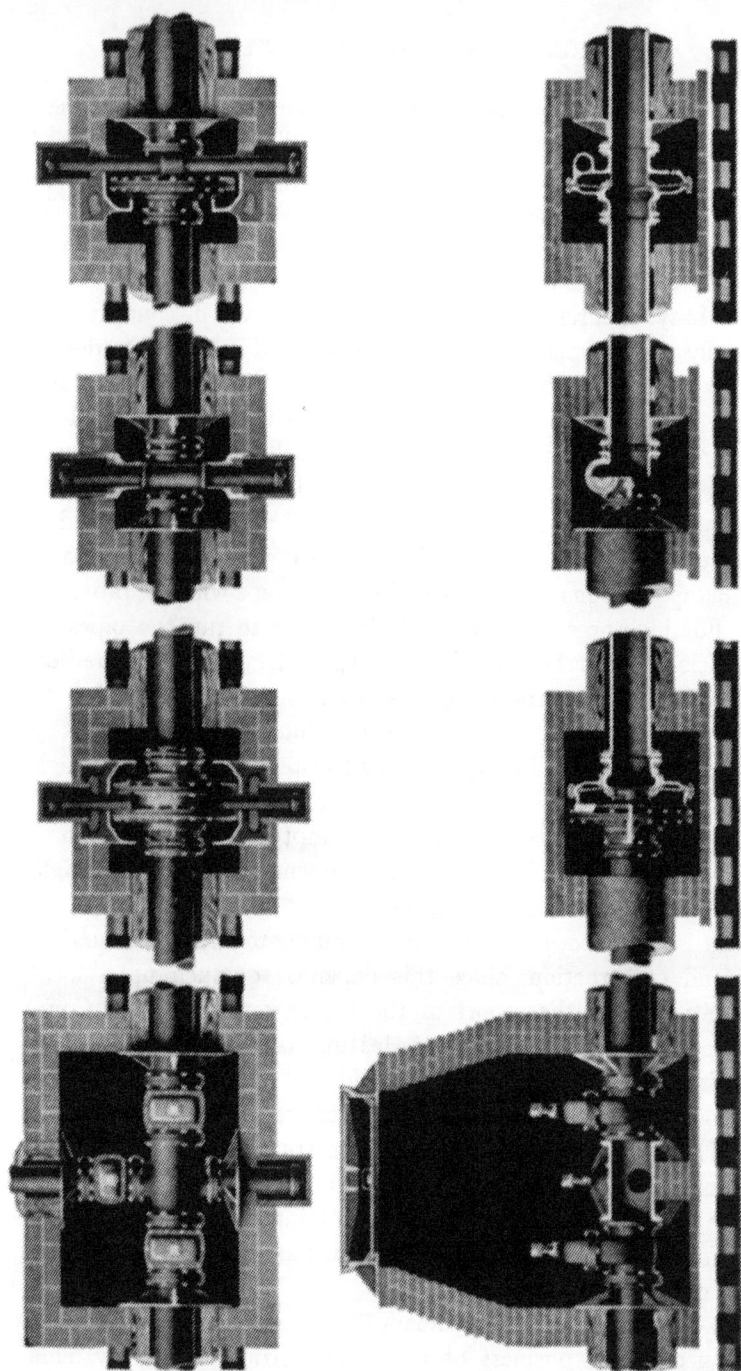

FIG. 16.—Details of underground construction and fittings

adopt standard forms of contract. The form given is similar to one used by the largest contractor in this line of work and appears to be equitable to both parties. Fig. 16 shows the

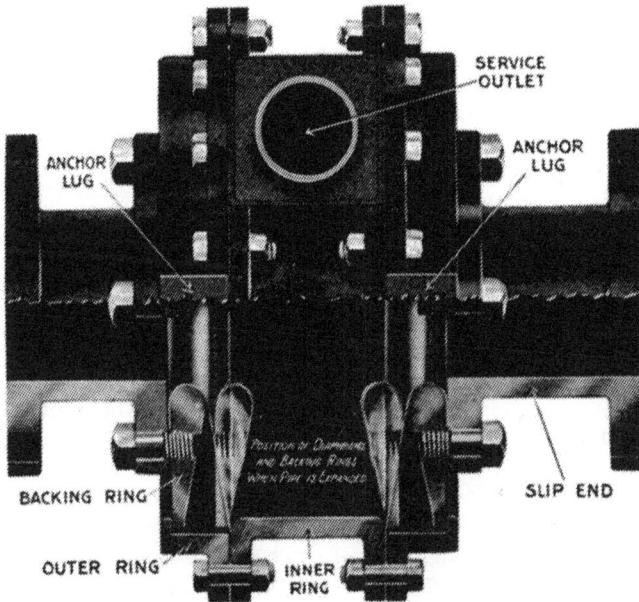

FIG. 16A —Variator type-expansion joint with copper diaphragms

FIG. 16B.—Double-expansion joint

general construction of fittings used, these illustrations being selected at random from the catalogues of the most prominent makers. While this form of contract may be considered a typical form, the typical clauses should be modified in a good many cases to conform with local conditions, and for hot-water systems the specifications must be altered to correspond with this type of construction.

GENERAL CONDITIONS

" General: 1. These General Conditions form part of the Contract and Specifications hereto attached.

Definitions of Terms: 2. In construing the Contract, the General Conditions and the Specifications, the following words shall have the meanings herein assigned to them:

" The ' Contractors ' shall mean the.... . ...and shall include its successors and assigns.

" The ' Purchasers ' shall mean........................ and shall include its successors and assigns.

" Location: 3. The underground steam-distribution system to be provided by the Contractors shall be located in the City of......................as indicated on the Contractors drawing No.......and detailed in ' Location and Sizes of Steam Mains ' Clause No. 36 of the specifications.

" The Power House of the Purchasers from which the steam is to be taken for distribution is located at.... ....

" Supervising Engineer: 4. The Contractors shall provide a competent and experienced engineer who shall at all times superintend the installation of the system and exercise a direct supervision of all parts of the work as it progresses.

" Skilled Workmen: 5. All skilled workmen employed in the actual work of installing the underground system of mains or having charge of any branch of the work shall be experienced men regularly in the employ of the Contractors.

" Patent Rights: 6. The entire system shall be installed in accordance with the latest approved form of construction adopted and introduced by the Contractors who are the owners of patents covering various component parts of the system.

" The Contractors shall at all times fully indemnify the Purchasers against any and every action, claim or demand, cost or expense or resulting damage arising from or incurred by reason of any infringement of patents in respect to any of the materials, apparatus or devices supplied by the Contractors under this contract.

" In the event of any claims being made or action brought against the Purchasers in respect to any alleged infringement of patents, the Purchasers shall immediately notify the Contractors thereof in writing and the Contractors shall have the right to conduct all negotiations or litigations which may arise therefrom.

" Provided by Contractors: 7. Except as herein otherwise specified, the Contractors shall supply all Pipe, Fittings, and materials and all Engineering, skilled and common labor, Cartage, Tools, Tackle and Plant of every description, necessary to carry on the work and complete the installation in an efficient and satisfactory manner. The Contractors shall provide all Masonry for the conduit and for the Boxes surrounding the Variators, Anchor Specials, Crosses and Fittings and for the Manholes for Valves, Expansion Joints and Traps, and all Drainage Tile required within the trench.

" Provided by Purchasers: 8. The Purchasers shall provide and maintain in good order until the steam mains are installed an Open Trench, free from all obstructions which would interfere with the installation of the work the trench to be constructed under the direction of and to conform to the grade determined by the Contractor's engineer.

" The Purchasers shall provide in the trench, all Crushed Stone or Coarse-Screened Gravel required and shall make all Outlet Connections for under-drainage from the drain tiles in the trench to the final points of discharge, and shall also make whatever provision is necessary to prevent leaks from water pipes, sewer pipes or other sources causing any accumulation of water in the trench.

" The Purchasers shall obtain from the proper authorities, all Permits which may be necessary for opening the streets, avenues, alleys, and public places, tapping sewers, changing piping or other purposes and shall at all times furnish to the representatives and employees of the Contractors the right-of-way or free access to all parts of the trench in which the mains are to be installed. The Purchasers shall properly guard and protect all parts of the trench and all material and apparatus on the streets and shall be responsible for all damages to any individuals or to the city caused by the trench or by the material or apparatus about it.

" When the steam mains have been installed complete in position, the Purchasers shall backfill the trench and repair and repave the surface where necessary, and remove all surplus material remaining from the excavation.

" Delays in Providing Trench: 9. If the Purchasers shall fail to provide the several sections of the trench complete as required and at such times as they are called for by the Contractors, or if the Purchasers shall fail to prosecute with diligence the work of preparing the trench, or any part of it, the Contractors shall have the right, upon three days' notice in writing to the Purchasers to proceed with the construction of the trench and complete it at the expense of the Purchasers; the intention being that the trench shall be prepared at such times and the work performed therein in such a manner as not to delay the work of the Contractors.

" Delays: 10. The Contractors shall be at liberty to leave any section or sections of the steam mains in an unfinished condition temporarily whenever, it shall be necessary for the purpose of making tests of the line or other sufficient reason, but as soon as the cause of the delay shall cease to exist, the Contractors shall complete their work on such sections of the line.

" Order and Precedence of Work: 11. In general the work of installation shall be started at that part of the system nearest to the power-house of the Purchasers and proceed thence to that part of the system most remote from the power-house, but whenever material cannot be conveniently assembled for the installation of the work in any particular section of the line and material is available for the installation of other sections, the work on such sections shall be given precedence. The Contractor shall have the right to carry on work at more than one location at the same time.

"Title to Installed Work: 12. All material, apparatus, tools and tackle furnished by the Contractors for use on the work included in the contract shall from the time of being brought to the location where the work is to be installed, be the property of the Contractors, whether installed in position, affixed to the freehold, or otherwise, until the final payment shall have been made of all amounts due under this contract, and the Contractors shall have the right, after three months' default in any of the payments herein provided, to take immediate possession of all the material and apparatus supplied under this contract, whether installed or otherwise, and of all the system of piping in which said material shall have been used including all rights to operate such system, and at the option of the Contractors to hold or dispose of it to the best advantage, applying the proceeds to the payment of any amounts due and unpaid under the contract, including on such unpaid amounts from the date on which they were due and in payment of all reasonable expenses incurred in the removal of such material or work installed, or the Contractors, shall, at their option have the right to operate the heating plant.

" In the event that the Contractors shall exercise their right to take possession of the heating system, and to operate it, the Purchasers upon giving thirty days' notice in writing to the Contractors of such intention, shall have the right to pay to the Contractors all amounts due and unpaid under the Contract, together with all expenses incurred by the Contractors in taking possession of and operating the system, and upon such payment being made the Contractors shall surrender to the Purchasers full possession of the heating system and of all material and apparatus provided and installed under this contract.

" At any time prior to the shipment by the Contractors of any material included in this contract, the Purchasers shall have the right to furnish to the Contractors the bond of a Surety Company satisfactory to the Contractors for an amount equal to the contract price and upon the receipt of such bond the Contractors shall waive their rights under this contract or to take possession of any material supplied or work installed under the contract or to take possession of and operate the heating system.

" Extra Work: 13.   No apparatus, material or labor except what is required to install the work included in this contract and the specification, shall be provided by the Contractors except on the written order of the Purchasers.

" Tests: 14.   The Contractors shall, in the presence of a representative of the Purchasers, test all steam mains at a pressure not exceeding forty pounds per square inch.   The test shall be made upon the completion of the work or by open tests of separate sections of the line of such convenient lengths as may be determined by the Contractors.   The tests shall continue in each case for a period of two hours and all work so tested and found right shall be accepted by the Purchasers at the time it is tested.

" In all cases where the Contractors are ready to test any sections of the line and it shall be found necessary to delay making the tests, owing to conditions over which the Contractors have no control, such sections shall be accepted by the Purchasers without test if those conditions shall continue for a period of three days.

" All steam required for testing the mains shall be provided by the Purchasers without charge.

" Measurements: 15.   All measurements of the steam mains shall be made on the surface of the mains over-all, including

Variators, Expansion Joints, Anchor Specials, Valves and all Fittings installed. Payments per lineal foot shall be made in accordance with those measurements and in addition to the prices named in the contract for Valves, Special Fittings and Street Traps.

" All measurements of the mains shall be made after they are placed in position and before the backfilling is commenced. Scale measurements from maps or plans shall be taken for general guidance of the work and for no other purpose.

" Guarantee: 16. The Contractors guarantee the successful mechanical operation when completed, of all the work provided installed by them under this contract and further guarantee that all work provided and installed by them under this contract shall be free from defects in material and workmanship for a period of one year from the date when the steam mains shall have been tested and found tight.

SPECIFICATIONS

" General: 17. These specifications form part of the Contract and General conditions hereto attached and describe and illustrate the material to be supplied and the work to be performed.

" Insulation: 18. All Pipes, Fittings and Connections shall be thoroughly covered and insulated against undue loss of heat from radiation according to the best standard practice.

" The Wrought-Iron Line Pipe shall be covered with a special insulating covering built up in sections and of such thickness as may be required to properly insulate the line.

" On lines operated under a normal pressure of twenty-five pounds or less the covering shall be two inches in thickness.

" On lines operated under a normal pressure above twenty-five pounds the covering shall be three inches in thickness.

" All covering three inches thick shall be installed in sections with " broken joints."

" The covering required under these specifications shall be . . . . . . inches thick.

" Immediately over the insulating covering there shall be applied two layers of one-sixteenth inch heavily asphalted waterproofing, each layer shall be applied in broken-joint method with sufficient lap for sealing and all laps shall be thoroughly cemented with liquid asphalt, or other material equally as good.

" A coating of No. 1 black waterproof varnish or other

material of equivalent value shall then be applied to the outer surface.

" Conduit: 19. After the sectional covering has been applied and waterproofed, a conduit shall be installed around the lines according to the following plans and specifications:

Description of Conduit

" At the proper height as specified in ' Underdrainage,' there shall be installed a continuous concrete base, four inches in thickness, forming the lower member of the conduit for the entire system. The concrete shall be thoroughly mixed in an approved manner and carefully installed in forms set to the correct grade by the Contractor's engineer. The upper surface of the concrete base shall then be cleaned and the conduit constructed on it in the following manner:. . . . .

. . . .

" Chairs, Rollers, and Saddles: 20. The pipes forming the steam mains shall be carried on cast-iron chairs, spaced at a distance not exceeding twelve feet. To insure in its proper alignment a free movement of the pipe, resulting from expansion or contraction, suitable rollers with cast-iron saddles or roller plates shall be placed between the cast-iron chairs and the sectional covering of the pipe.

Wrought-iron Line-pipe: 21. All Wrought-iron Pipe used in the installation shall be strictly wrought-iron line-pipe fitted with long recessed couplings or plain ends, if joints are to be welded, and shall be of the highest quality obtainable. Each length of pipe to be tested by the manufacturer before shipment and found tight under a hydraulic pressure of 500 pounds per square inch. The pipe shall be of the following standard weights per lineal foot with the usual allowance of $7\frac{1}{2}\%$ for variation in rolling etc.

" Couplings and Flanges: 22. At all joints in the pipes except those adjacent to special fittings, the Couplings shall be of the heavy long pattern recessed type, except where the joints are welded. All joints connecting fittings other than Variators shall be made with Standard Flanges accurately faced and drilled and provided with soft annealed corrugated copper ring gaskets.

" Cast-iron Pipe: 23. Any Cast-iron Pipe supplied shall be of the proper thickness for the pressure under which it is to be operated. All such pipe shall be of the best quality and workmanship, with flanges cast integral with the pipe, or fitted with threaded ends.

" Fittings: 24. All Anchor Specials, Angle Joints and all Special Fittings, Crosses, Tees and Elbows shall be flanged and accurately faced and drilled. All crosses, tees and elbows shall be provided with sweep angles and shall have eccentric openings at all reducing branches to permit a free discharge of all condensation.

" Valves: 25. Flanged Gate-valves of the packingless type, specially designed and manufactured for this class of construction shall be used in the steam mains throughout the work. All main Valves shall have double gates with stems supplied with square nuts of uniform size, or if preferred by the Purchasers the valves will be fitted with hand wheels.

" All valves used in street construction shall be bolted directly to the street corner specials and where two or more valves are located at the same corner, the valves shall be bolted to the same special and made conveniently accessible in one manhole. Where the conditions are such as to make it advisable, reducing flanges and nipples between the street corner specials and the valves may be supplied.

" Variators and Expansion Joints: 26. Provisions for all variations in the length of the steam mains resulting from the expansion or contraction of the pipe, caused by changes in temperature, shall be made at proper distance in the line either by the packingless Variators of a standard type or by an improved form of brass slip Expansion Joints.

" When Variator Construction is installed the distance between Anchor Fittings and Variators shall not be greater than fifty-one feet.

" When the steam mains are installed in sections where there are occasional buildings only, requiring a steam service, the distance between Anchor Fittings and Expansion Joints shall not be greater than 120 to 260 feet, varying according to the size of the pipe.

" All variations and Expansion Joints shall have outlets in their stationary sections to permit service connections being taken from fixed points and at a level either above or below the outside diameter of the pipe and at right angles to the line of steam mains.

" Wherever Expansion Joints requiring packing are installed, manholes shall be provided.

" Anchorage and Anchor Devices: 27. The entire steam-distributing system shall be securely anchored at suitable points in the line.

" All Anchor Specials and all street corner specials and all single expansion devices shall be securely anchored to the adjacent brick, concrete walls or base.

" Cast-iron collars shall be provided in connection with the anchorage between the iron pipe and the conduit.

" All Anchor Specials shall have outlets for service connections located at a level either above or below the outside diameter of the pipe and at right angles to the fitting.

" All changes in the grade or in the alignment of the steam mains shall be made at anchorage points.

" Service Connections: 28. All service connections shall be taken from fixed points in the line and to insure that dry steam shall be delivered from the mains, the service connections shall be taken from above the upper line of the steam mains excepting that wherever it may be necessary for draining the mains, the service connections may be taken from the underside of the fitting.

" Provisions for future service connections shall be made by means of nipples of sufficient length to extend through the masonry. The nipples shall be closed with cast-iron caps and protected against corrosion.

" Street Traps: 29. Unless otherwise provided, the water of condensation shall be taken from the street mains at all low points, by automatic low-pressure Steam Traps, equipped with long pattern automatic air-valves, and located in independent manholes. The street steam-traps shall be located with the inlet below the steam mains and discharge through vitrified sewer-pipe into the city sewerage system, or to a point where the condensation may be conveniently disposed of or utilized. A " U " trap shall be placed between the steam trap discharge and the sewer connection to prevent sewer gases entering the manhole. The steam trap shall be by-passed in such a manner that at all times any accumulation of water of condensation in in the steam main may be taken care of, whether the steam-trap is in operation or not. Valves shall be placed ahead of the steam-traps and in the by-pass so as to permit the steam-trap to be removed while steam is in the street mains and to permit the operation of by-pass either independently or in connection with the street traps. A tell-tale pipe shall be so placed that the steam will escape to the atmosphere whenever the trap is " blowing." or when the steam is passing through the by-pass.

" Screwed cast-iron fittings. wrought-iron pipe and screwed brass gate-valves shall be used.

" When conditions make it impracticable to place the street trap in an independent manhole, the trap may be placed in the same manhole with a valve or an expansion joint. In such a case the manhole shall be enlarged.

" Wherever possible provision shall be made in the construction of the trap manhole, and in placing the trap, to permit of the installation of a street condensation-meter.

" Brick, Concrete and Tile Work: 30. All Variators, Anchor Specials, Angle Joints and Special Fittings shall be enclosed in boxes constructed of Concrete, hard burned brick or hollow tile.

" The side walls of the boxes shall be carefully laid with cement mortar and coated with cement. The boxes shall be built on a bottom course or flooring of concrete or brick carefully laid with cement mortar. The top of the box shall be covered with suitable planking. The top of the planking shall be covered with three-ply tar paper, and one layer of brick or tile shall be carefully laid on the tar paper and thoroughly grouted.

" Standard grades only of American Portland Cement shall be used.

" Manholes: 31. All manholes shall have an inside diameter of not less than eighteen inches and where two or more valves are placed in the same manhole the inside diameter of the opening shall be of sufficient size to permit the valves being operated by extension wrenches.

" All manholes shall have cast-iron, double cover curbs with the inside cover packed to prevent water entering from the surface of the street. The outside cover shall be circular in form and of sufficient strength to carry safely a load of five tons. The mason work for the manholes shall be constructed as specified in clause No. 30.

" Grading of Mains: 32. All steam mains shall be laid to a suitable grade to be determined by the Engineer of the Contractors and all mains between Anchor Fittings, Variators or Expansion Joints shall be laid in a straight line. Wherever obstructions are encountered or the natural conditions are such as to necessitate dropping special points in the mains below the regular grade, provision shall be made for suitable trap connections to automatically relieve the mains of the water of condensation.

" Underdrainage: 33. Drain tile shall be laid underneath all the steam mains, special fittings and valves forming the steam-distribution system. to drain off any water due to seepage, infiltration or leaks from sewer connections or water mains.

The drain tile shall be not less than four inches in diameter laid to the proper grade and connected to the City sewerage system or other suitable outlet wherever necessary to insure proper drainage of the system. The upper side of the drain-tile shall be located at a level of not less than nine inches below the underside of the conduit system for Variator Construction and not less than three inches for Expansion Joint Construction.

" In all parts of the distribution system where the steam mains are of a larger diameter than five inches, two lines of drain tile shall be installed. The Purchasers shall provide the trench and provide and lay all special tile or sewer pipe required to reach a suitable point of discharge not adjacent to steam mains.

" The Purchasers shall supply in the trench whatever crushed stone or coarse screened gravel may be required to fill in and around the drain tile and under the conduit. The crushed stone or gravel should be carried up to the top of the conduit unless the soil in which the trench is excavated is of such a character as to render the use of tile stone, or gravel unnecessary.

" Trenches, Backfilling and Water Leaks: 34. All labor and material required in excavating, guarding, backfilling, repairing and repaving the surface of the trench in which the underground steam mains are laid shall be provided by the Purchasers. All trenches and excavations shall be backfilled with dry earth, thoroughly tamped to a level of not less than one foot above the top of the Conduit and no puddling shall be allowed in any of the backfilling.

' Before backfilling is commenced all leaks from water pipes, sewer pipes or other sources shall be remedied by the Purchasers who shall take proper precaution to guard against a recurrence of such leaks from the same or similar source.

" Special: 35. In all sections of the steam-distribution system where the steam mains have an outside diameter of eighteen inches and over, the conduits in which the steam mains are to be installed shall be constructed with the tiles forming the roof of the conduit built up in arch form.

" Location and Sizes of Steam Mains: 36. Following is a list of the Streets, Avenues, Alleys, and Public Places in which Steam Mains are to be installed and the estimated length of each size of pipe to be laid.................................

The attached drawing also includes the approximate location of pipe-lines, manholes, expansion joints, etc., and is made a part of the specifications.

(*b*) METHODS OF PIPING WITHIN BUILDINGS

When a heating company enters the field, it usually discovers a great many different classes of piping in customers' installations. It is often necessary before service can be g ven that extensive changes be made in the piping arrangements. In many cases, this is due to the fact that the p ping is not arranged so as to provide circulation on the comparatively low pressures which it is desired to carry on the distribution lines. In large cities where high pressure lines are used, this phase of the problem is not so vital to the company, although it is often to the interest of the consumer to rearrange his pipe-lines in such a way as to realize all the economies obtainable. In many smaller towns the companies refuse to supply a consumer unless his system meets certain definite requirements, and some have gone so far as to demand that all new customers install the latest forms of atmospheric and vapor systems all with the idea of reducing the back pressure on their lines. These precautions are especially necessary where flat rates are in effect, since some method of heat regulation is manifestly the sole protection that the company has against waste of heat. It is impossible with some of the old-fashioned systems to keep the consumption of steam within reasonable limits and satisfactory service is only obtainable after a thorough overhauling of the piping and radiation.

It is said that there are a hundred or more vapor, atmospheric, or vacuum systems which have been patented and put on the market. A brief description of the essential features of each distinct type will probably suffice for the understanding of the different systems encountered. The following tabulation will show the methods of distribution of steam systems.

There are three methods of steam distribution; Direct Heating (Non-ventilating), Direct-Indirect Heating (Semi-ventilating) and Indirect Heating (Positive ventilating), which may be tabulated as follows:

DIRECT HEATING.

Up-feed   Down-feed

One-pipe   Combination,   Two-pipe

Gravity System   Vacuum System.

(1) Direct return or loop   (a) Pump for removing water and
to boiler.   air accelerating returns
(2) Return through trap or   (b) Ejector for removing air from
receiver before enter-   radiators
ing boiler   (c) Mercury seal

Low Pressure.   Atmospheric System.
Medium Pressure.   Vapor System
High Pressure.

DIRECT-INDIRECT HEATING.

Natural air draft through wall
opening over direct radiators.

INDIRECT HEATING.

Mechanical air draft over hot
blast coils or radiators, with
or without air washer.

Plenum or pressure   Exhaust or suction.
(Disc propeller.   (Paddle wheel or
blade fan)   multiple blade fan)

Steam engine or electric
motor drive.

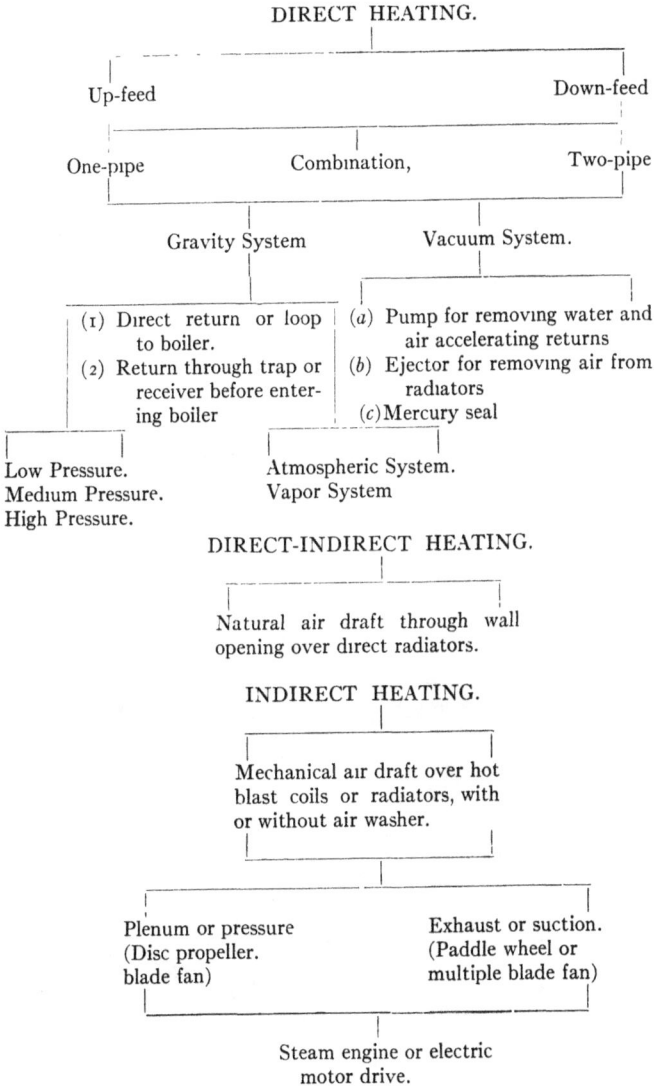

Figs. 17 to 20 illustrate the arrangement of some of the
various types of piping above mentioned. Fig. 17 (a) shows the
common method usually found in small steam-heated buildings.
It consists merely of a main distributing-pipe near the ceiling
of the basement from which risers are tapped out where needed

for the radiators above. This system is the one-pipe gravity
type in which the steam supply and the returning condensation
are flowing in one pipe, generally in opposite directions.   If the

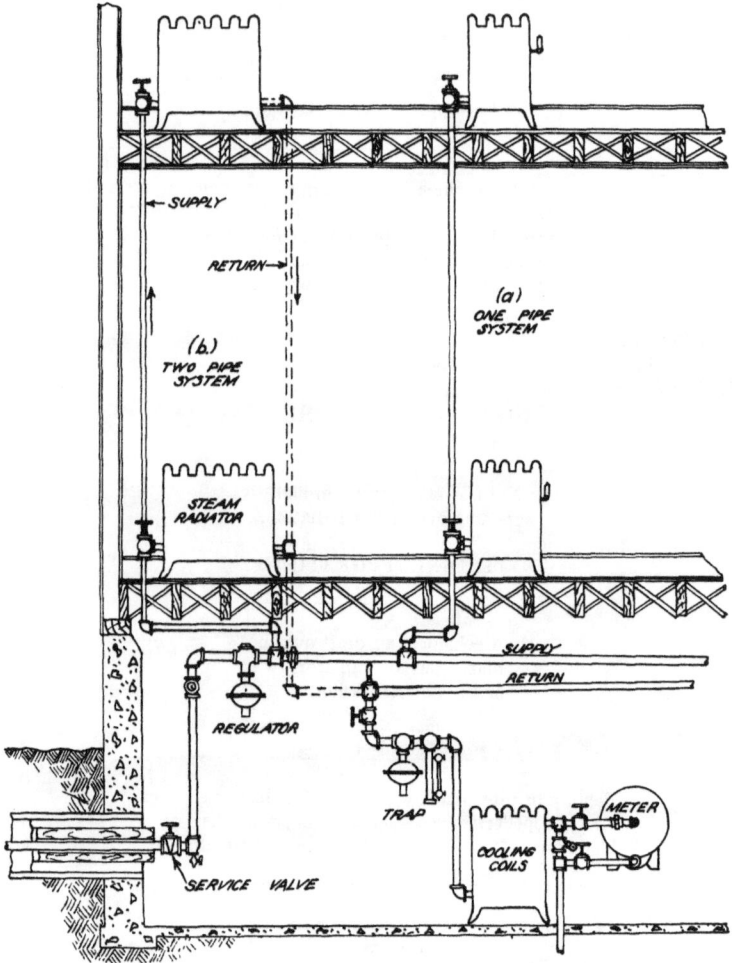

Fig. 17.—One or two-pipe gravity steam-heating system

pipes are large enough this system will be satisfactory provided
proper pitch is given in all parts of the system.   Any sagging
or trapping in the lines will require excess pressure for lifting

the return water up to the level of the return piping. The radiator supply is taken through a hand valve at the bottom, and there is no provision for removing the entrained air except through the automatic air vent on the opposite end of the radiator coil. These valves or vents frequently get out of adjustment and as a consequence the radiator becomes air bound, thus stopping the circulation of steam and the result is a cold radiator and a complaint lodged with the heating company. Obviously, repeated instances of this will lead the company to seek a better system if possible.

The first improvement in this class of installation was brought about by the addition of an independent pipe for bringing the returns back from the radiators. Thus we have the two-pipe gravity, up-feed system, Fig. 17 (b). Fig. 18 (a) shows another method of returning the condensation. This is known as the one-pipe, overhead or down-feed type, and it will be seen that the objectionable feature of the one-pipe system—the returns forcing their way back against the flow of steam—is avoided. This arrangement is further modified by using separate return lines, viz. the two-pipe overhead system. Two-pipe systems are also instrumental in relieving the system of air, and are much to be preferred for this reason if no other.

The next development consisted in the attachment of vacuum or exhaust pumps to the returns in one of two ways. The Paul system consists of an independent exhaust line fitted onto the radiator as shown in Fig. 18(b). This is run to an ejector or some apparatus in the basement which relieves the radiator of air through an automatic expansion air-valve, inducing free circulation and allowing the return water to flow back in the same manner as in the one or two-pipe gravity type. The Dunham, Webster, Illinois Engineering, Monash, and other vacuum systems are all of the two-pipe up or down-feed type, see Fig. 18(c). The returns—water and air—are removed by action of a vacuum pump in the basement, each radiator being equipped with a separate return valve. The thermostatic valve is constructed so that it will open up into the return when the temperature within the radiator decreases below a certain limit,

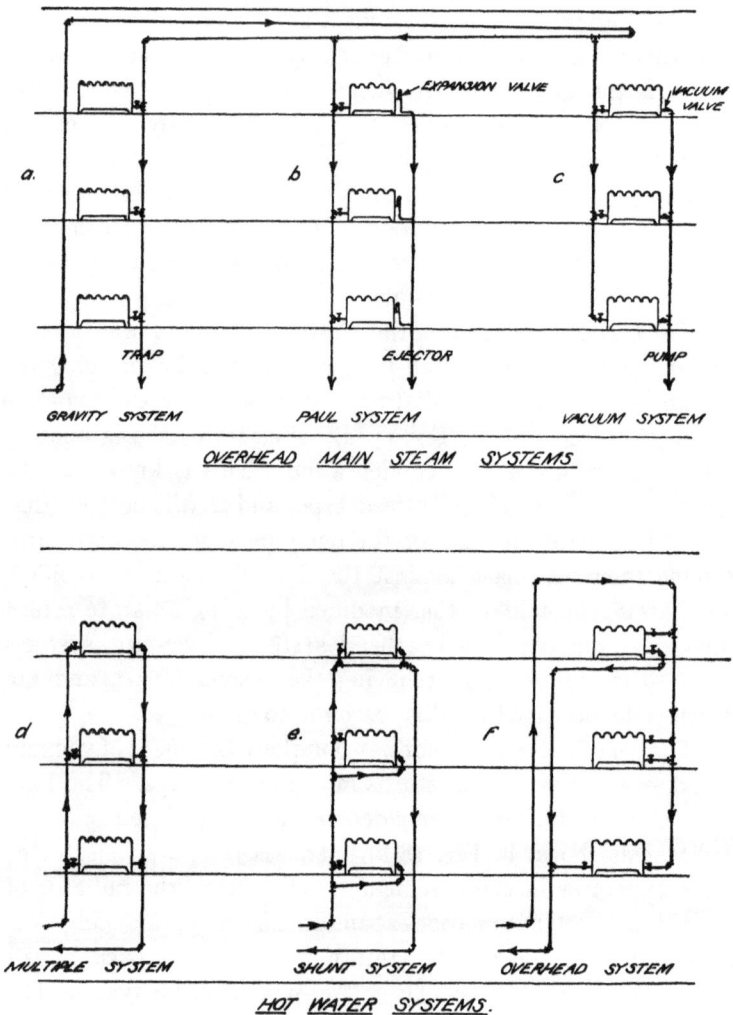

FIG. 18.—(a) Overhead, single-pipe steam-heating system
(b) Paul system
(c) Vacuum return system
(d) Hot-water piping diagram—Multiple system
(e) Hot-water piping diagram—Shunt system
(f) Hot-water piping diagram—Overhead system

the suction of the vacuum pump removing same as fast as the valve permits. The inrush of steam into the radiator following the removal of air and water again expands the valve and closes it when the temperature is regained. The float valve or float trap serves the same function as the thermostatic valve or trap, but its action depends upon the water condensed in the radiator to lift the float and allow the water to be drawn out by vacuum. This vacuum system is installed only in large buildings and with proper attention to design it is possible to circulate the steam at only a few ounces or practically atmospheric pressure.

Fig. 19 illustrates the Vapor system of heating. The arrangement is a two-pipe system, hot-water radiation being used. Hot-water radiators are constructed so that the separate loops or sections are joined together at both top and bottom, by pipe nipples which allow free circulation in all parts of the radiator; (see Fig. 20), whereas the steam-type radiator does not have the top of the coils so joined, the only transverse opening being at the bottom. In other words, the loops or sections of steam radiators are merely inverted chambers with no opportunity for the steam to flow from one section to the adjoining one, except at the bottom. In the Vapor system the radiator inlet-valve is tapped in near the top and is of carefully computed size for the conditions prevailing. The returns are brought back to the basement where they are sealed, vented to the atmosphere and trapped into a meter or sewer line. No air valves are required on the individual radiators, the only fitting on the return line being an ell, which relieves the radiator both of air and water.

The Atmospheric system, Fig. 20, combines many of the features of the Vapor system just described. Steam is delivered through a regulating valve on the street service line, where it enters the basement and is reduced to about eight ounces pressure. The piping to the radiators must be of sufficient size to allow the circulation with practically no drop in pressure. Extra heating surface of the hot-water type is provided so that the water of condensation has a chance to cool before entering the return pipe. No valve is used on the return end of the

radiator, all the water flowing back by gravity to a pipe-seal in
the basement. This seal has three outlets; the inlet line; vent
line which is carried up about 10 to 20 ft. above the seal or

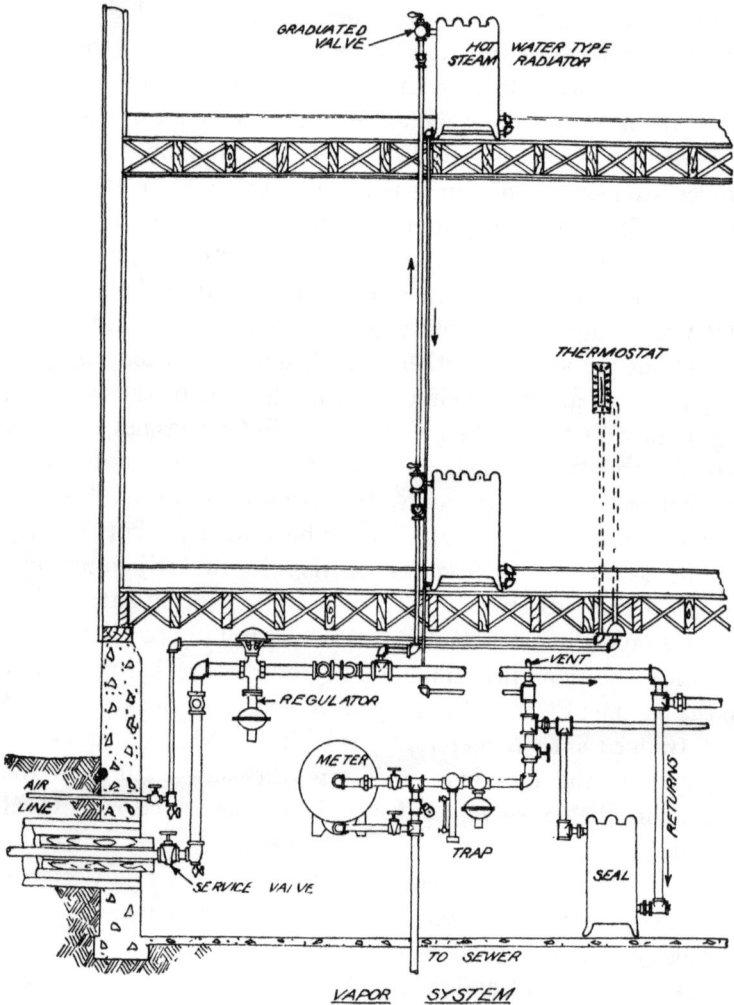

FIG. 19.—Vapor system of heating

receiver; and the outlet pipe to the meter. This system is used
with great success in many classes of buildings, although it
has not yet displaced the vacuum system for high buildings.

The Atmospheric system is very easily controlled. By means of the graduated inlet-valve any desired percentage of the heating surface is brought into action by hand control, and the thermostatic feature is therefore omitted. Regulating or throttling the flow of steam to the ordinary radiator does not

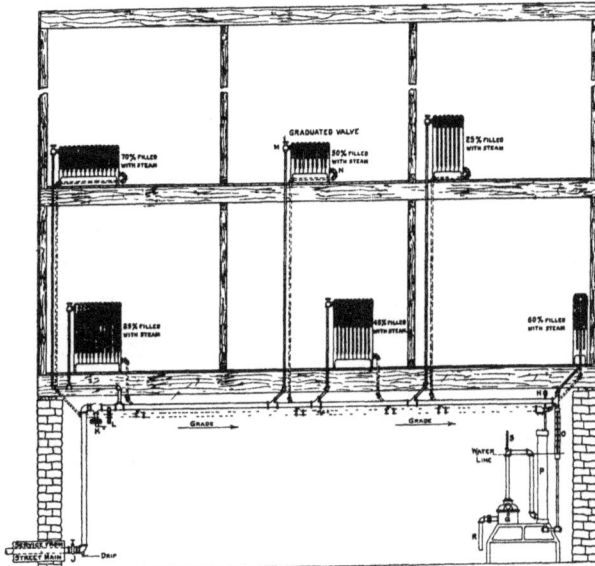

THE ATMOSPHERIC SYSTEM OF STEAM HEATING
CENTRAL STATION SUPPLY
'ADSCO' SPECIALTIES

FIG. 20.—Atmospheric system of steam heating

prevent the entire radiator from being filled provided the air is exhausted, but when the hot-water type radiator is used, the steam enters at the top and being lighter than the air remains there until condensed. This feature is made clear in Fig. 20. The principal economy of the Atmospheric system lies in the low temperature to which the water is cooled, every possible heat unit being usefully expended.

Hot-water heating systems are not essentially different from steam systems, but a different type of radiator is used as has just been seen.   Fig. 18(*d*), (*e*) and (*f*) illustrates the methods of connecting the system to the central station street service. A regulating valve is used to control the quantity of water circulated and this may be equipped also with a thermostat which shuts off the supply when the building is sufficiently heated. Without this valve the customers nearest the plant obtain the higher differential pressures and consequently an excessive amount of water flows through their radiators without accomplishing useful results, and necessitating increased cost to the company for pumping.   The financial success of hot-water heating systems depends upon the attention given to details pertaining to the customers' installations with a view to reducing the pounds of water per square foot of surface to a minimum, consistent with the drop in temperature, which will average around 30 degrees.

Hot-water heating customers may be attached to steam-distribution systems by means of a device called the transformer. This device is nothing more than an efficient hot-water heater which serves the same functions as a fuel heater.   The steam enters the tubes of the transformer and heats the water which circulates by gravity, due to difference in temperatures—the hot water rising upward and the cooled water circulating downward, to be reheated.   The condensed steam is metered as before described.

The automatic regulation or control of the heat supply is accomplished by one of two methods, viz.: thermostats actuated by air pressure supplied through an independent compressed air line from the heating station; and automatic self contained devices which do not depend upon any outside agency for their operation.   The former devices are losing favor because of the difficulties found in maintaining the underground air line.

For hot-water customers and flat-rate steam customers, the company should insist upon the best type of equipment being installed for regulation of the temperature.   Thermostats require a great deal of care and adjustment and unless proper attention is accorded them, they become inefficient and fail to serve the purpose intended.

The various diagrams show also the method of installing condensation meters on the premises and a further discussion of this phase of the subject will be found on p. 147, Chapter IV. It is highly desirable to reduce the temperature of the condensation to the lowest point possible for the following reasons: First, the customer is entitled to the benefit of all heat which can be abstracted from the condensation, and second, the company will find satisfactory meter performance much enhanced by the same means. Many companies require economizer or cooling coils to be inserted between the trap discharge and the meter. This economy coil is merely an indirect radiator, a current of air being heated by the hot water. The size of this coil is usually based on 20% of the connected radiation. This scheme is open to various objections. Where domestic hot-water heaters are needed the returns may be circulated through the tubes of these heaters with resultant economy.

CHAPTER IV

METERING

INTRODUCTION. (I) WATER METERS. (II) STEAM METERS. (III) CONDENSATION METERS. ACCURACY OF METERS. RULES FOR CARE OF METERS. GENERAL CONCLUSIONS.

No other factor is contributing more to the present progressive state of the heating industry, or promising more for its continued prosperity than that of metering the heat supply. Those early pioneers who first launched this industry evidently possessed sufficient foresight to perceive the essential dependence of its ultimate growth and extension upon some such basis as metering. Accordingly we learn that the first inventions of steam-measuring devices were contemporaneous with or immediately antecedent to the first systems of heat distribution installed.

Indeed, the first available records point to the fact that the father of the steam meter was Birdsill Holly, who as noted in Chapter I, worked out a system which still remains, with various modifications, the basis of the industry as we find it today. Appreciating the fact that it would be more equitable to both company and consumer if the quantity of service could be measured, he made application November 29, 1880 for a patent on meters, which was granted May 10, 1881.

While it is barely possible, as so frequently happens, that some other inventor was at work simultaneously upon a similar device, no record of such appears, and to Mr. Holly is due the credit for consummating his idea by the actual construction of this meter. It is interesting to note here that in all probability this invention antedates by a few years the first Edison electric chemical meter, which was used quite extensively in the early days of incandescent lighting. This furnishes another instance pointing to the priority of the introduction of central heating over its closely allied industry, central electrical systems.

Briefly, the essential features covered in the Holly patent are as follows:

" This invention relates to improvements in meters for steam or other fluids, in which the paddle-wheel is rotated by the impact of steam or other fluid, and the number of its rotations counted and recorded by a counting and recording mechanism similar to the mechanism in gas meters, the steam or other fluid being guided to the paddle-wheel by a self-adjusting guide and the movement or rotation of the paddle-wheel regulated or controlled by the application of a hydraulic brake as hereinafter described."

After the appearance of the Holly meter, Figs. 32 and 33, a large number of which were manufactured and used with varying success by a few companies, it seems that no great amount of attention was devoted to the improvement of such devices for a period of more than a decade. During this interval the industry encountered a lull in activity as far as metering is concerned, those stations which were then installed being forced through lack of a really dependable meter to adopt various methods of flat-rate charging. With the appearance of the condensation meter in 1904, a renewed impetus was imparted to the scientific development of commercial heating. It is probably on account of inability to obtain a satisfactory steam meter at this period, that heating companies were led to adopt the natural substitute—measurement of condensation, which system now enjoys a greater measure of popularity than ever before, notwithstanding the substantial progress which has been made in the design of steam flow meters.

While it would be of historical interest to trace the evolution of meter design from its earliest known inception, yet such a narrative would be of questionable value and would probably not appeal to practical readers. For this reason, this chapter will be mainly directed toward a study and description of modern practice, with a brief consideration of the fundamental principles and laws underlying the design of meters. The great development along all technical lines in this present age of science and research has led to a great and urgent demand for exact methods of measurement. Refined laboratory apparatus is often not adapted to every-day use; and inventors, perceiving the demand for a more practical form of instrument for steam metering, have turned their efforts toward perfecting really serviceable devices. These efforts followed divergent lines, and it is not surprising therefore, to learn of the great variety of attempts to solve the problem of measuring the output of heat-generating systems. Many efforts have ended in failure because it was frequently not recognized that meter design depends upon a very exact knowledge of the natural laws governing the flow of fluids; and even to-day inventors sometimes attempt to construct meters which ignore certain of these facts. Many of the faults which appeared in the first designs have been slowly eliminated by a process of experimentation until there are now on the market a few steam meters which may be considered dependable for commercial use.

The word " meter " as used in this chapter includes all methods of measurement although, technically considered, many of the devices mentioned are not properly meters in themselves, but fulfill only in part, the strict definition of the term. A meter is '' an instrument which indicates, records and integrates automatically the quantity, force or pressure of a fluid passing through it or actuating it." A fluid is a substance capable of flowing and may be in either of the following states:

| Fluids. | Inelastic. | Liquids, which expand, only by evaporation, and are relatively incompressible. | Water. Oils and various industrial substances. |
| | Elastic. | Gases or vapors, which expand or compress while retaining their original homogeneity of structure. | Gas, Air, Steam. |

It is of course improper to consider a weir, for example as a meter, since the weir, minus any secondary attachment for making a record of the variable flow is nothing more than a device employed by engineers to enable them to take visual measurements of the area of cross-section of the liquid, which measurements may be used in mathematical computations. This distinction must not be lost sight of, and accordingly in this treatise all automatic meters are considered to embrace two elements: the primary element (*a*), which is the motive or differential producing portion of the device, and which communicates its action to the secondary element (*b*), consisting of suitable indicating, recording or integrating mechanisms.

As a prelude to Tables I, II and III the nomenclature used may well be subject to explanation here. In Table III, steam meters are classified into two divisions, Shunt type and Series type. These terms refer respectively to those meters wherein (*a*) only a portion of the flowing fluid to be metered comes into actual contact with the primary element of the device and (*b*) where the entire volume of steam must pass through the same. The terms " Velocity Meters " and " Area Meters " are explained at length later on in this chapter. The terms " Primary " and " Secondary " elements have been explained in the previous paragraph.

There are three methods of measuring the output of a central steam station, and these are:

1. Water meters, which register the amount of water fed into the boilers.

2. Steam meters, which register the amount of steam delivered from the boilers.

3. Condensation meters, which register the amount of return waters of condensation from the various customers.

I. WATER METERS

Several methods are available for metering the boiler feed supply. These are:

a. *Displacement Meters*, consisting of: (1) Revolving vanes or discs, and (2) Reciprocating plungers or pistons, geared to dials which record the volume in suitable units: viz.: cubic feet or gallons.*

---

* Certain Area Steam meters, e g. the St. John, are also adapted for metering water.

b. *Velocity Meters*, operating on the well-known principle of the Pitot tube or Venturi tube and registering velocities through appropriate mechanical manometer devices. (See also " Velocity Steam Meters.")

TABLE I. CLASSIFICATIONS OF WATER METERS

	Primary Element.	Secondary Element	Maker, or Trade Name.
I **Displacement** **Meters**	Disc	Gears and Dial	Niagara * Keystone * National * Worthington *
	Vanes or Turbine	Gears and Dial	Worthington * Kreutzberg *
	Plungers or Pistons	Gears and Dial	Worthington * National *
II **Velocity** **Meters**	Pitot tube	Mercury Manometer	General-Electric *†‡ Simplex *†‡
	Venturi tube	Mercury Manometer	Venturi— Builders Iron Foundry*† Simplex—*†‡ Simplex Valve & Meter Company*†‡
III **Wier Meters**	"Yorke" Proportional Height Wier	Float	Precision— Precision Instrument Co.*†‡
	"V-Notch" Orifice	Float	Lea-Yarnall-Warring Co.*‡
IV **Automatic** **Weighers**	Gravity chambers or compartments	Floats or trips with counter	Weco-Weigher— Wilcox Engineering Co.* Hammond Meter— Alberger Pump & Condenser Company * Kennicott Weigher *
	Revolving Drum	Revolution Counter	Detroit Boiler Feed Meter— Central Station Steam Company *

* Integrating   † Indicating   ‡ Recording or Autographic.

c. *Weir Meters*, consisting of floats for automatically gauging the height of a stream of water flowing over or through various forms of outlet or orifice, e.g., the " V-Notch," " Yorke " and other forms of weirs.

d. *Gravity Automatic Weighers*, consisting of chambers or compartments which alternately fill and discharge, differing only in function from the condensation meter.

Space will not admit of an extended discussion of these various devices, and indeed, for the specialized demands of the commercial heating industry, they are not the goal, but rather the temporary stepping-stone to the more modern and desirable methods of measurement. Sufficient therefore will be a tabulated summary of the more important of these meters now to be obtained on the market.

FIG. 21.—Disc-type displacement water meter

The above outline as in other classifications to follow does not pretend to include all the devices now manufactured; but only those which are in more general commercial use today. Perhaps the reader can gain no better idea of these devices than by illustrations, accordingly, Figs. 21 to 24 are given. For a continued discussion, see also p. 154. (Limitations of Accuracy.)

These various devices are desirable for ascertaining records of station output, and for determining various factors, such as: evaporation, efficiency, rate of steaming, etc., and therefore some form of them should be found in every first-class steam-generating plant, if not equipped with steam meters. Without their agency the efficiency of the boilers is a matter of guess work and slipshod methods may be expected to prevail not only in the boiler room, but in other departments if no interest is taken in maintaining a complete plant record at all times.   The meters themselves should be frequently checked for accuracy by accu-

FIG. 22.—Venturi water and steam meter

rately weighing the water fed to the boilers.   When curve-drawing instruments are used they indicate not only the total amount of evaporation but also whether or not the water is fed into the boilers in a steady and economical manner.

The theories pertaining to several water meters are discussed under " Steam meters."

II. STEAM METERS

The difficulties attending the metering of steam for heating or similar purposes are only appreciated after a thorough study of the different factors involved.   In view of the apparent success of other forms of meters, gas, water and electricity—it would be expected that man's ingenuity could as easily overcome all

FIG. 23.—" V-notch " Weir water meter for boiler feed

obstacles to invention in the field of steam engineering.   We find, however, that the quantitative determination of steam flow involves a surprisingly large number of variables, affecting the practical limits of design and accuracy.

Could steam be depended upon to obey the theoretical laws of thermodynamics under all ordinary conditions the solution would be evident; but unfortunately steam cannot be considered a perfect gas, amenable to such formulæ.   Only when its temperature is maintained well above the point where any moisture can be present (that is to say superheated), will theory coincide with practice. This is seldom the case for any appreciable length of time, and in commercial work the substance is known to be extremely unstable and elastic in form, its properties subject to continuous alteration. The various conditions affecting the measurements are: Density, Pressure, Temperature, Moisture and Superheat, and while it has been the constant endeavor of inventors to compensate automatically for as many of these variables as possible, yet today, there is no steam meter on the

FIG. 24.—Detroit boiler-feed meter

market which is automatic in the strict sense applicable to gas, water and electric meters. By this is meant, that all steam meters require the addition of more or less computation for securing exact results.

Prof. G. F. Gebhardt is one of the pioneer investigators of the subject of steam meters and has contributed several valuable articles to the comparatively small amount of literature devoted to this important field of engineering.* In *Power*, Vol. 35

* Trans. A. S. M. E. vol. 31. See also J. A. Knesche, Engineering Magazine, Vol. 44 (Nos. 3 and 4).

(Nos. 6 and 7) he published two articles, the first one of which contained a classification of the steam meters which had been exploited up to that time (1912). This schedule has been drawn upon in the preparation of Table II, but the subject is treated somewhat differently. Under this table there are primarily two main types:

a. *Area Meters*, which include only the floating-valve type of Meters, and

b. *Velocity Meters*, which comprise all other meters.

TABLE II. CLASSIFICATION OF STEAM METERS

		Primary Element	Secondary Element	Inventor or Maker	
Area Meters (Velocity constant)	Series type	Floating valve	Mechanical control, clock and chart	St. John	1893*†
				Gehre	1896*†
				Baeyer	1902*†
				Bendemen	1902*
				Sargent	1908*
				Lindmark	*
				Rhenania	1910*†
Velocity Meters. (Area constant)	Shunt type	Impeller in current of steam	Gears and dials	Lindenheim	1896‡
				Gebhardt	1908‡
	Series type	Pitot tube	Mercury-manometer	General-Electric	1910*†‡
				Republic	1914*†‡
				Bristol	1912*†
			Water-manometer	Burnham	1906
				Gebhardt	1910*
		Stationary or throttling disc	Mercury-manometer	Halwachs	1907*†‡
				Gehre	1907*†
				Sarco	1910*†‡
				Storrer	1910*†
		Venturi tube	Bourdon-manometer	Eckhardts	1903*†
			Mercury-manometer	Bristol	1911*†
			Mercury-manometer	Parenty	1886*†
				Herschel	1910*†‡
		Impeller in current of steam	Gears and dials	Holly	1880‡

* Indicating.   † Recording or Autographic.   ‡ Integrating.

The Fundamental Laws of Design. Any attempt to present a discussion of meters would be incomplete without some reference to the mechanical laws of design, or the theoretical phases

of the problem. Therefore, several formulæ must be given in
order to gain an accurate knowledge of the fundamentals. An
effort will be made to avoid a severe mathematical treatment of
the subject and as will be seen, only elementary equations which
will be easily understood are employed. As the designation
implies, area meters are devices designed to determine the flow
of fluids by means of a varying cross-section or orifice. The
fundamental equation for the flow of any fluid is as follows:

$$W = A \times d \times V \times K \quad . \quad . \quad . \quad . \quad . \quad (1)$$

in which

W = Weight in pounds per second.

A = Area of orifice in square feet.

d = Unit density or weight of the fluid in pounds per cubic
foot.

V = Velocity in feet per second.

K = Coefficient determined by experiment.

Upon this formula is based the design of all types of meters,
including both area and velocity meters.

When certain known values are assigned to the various
factors, in equation (1) a solution is possible. Suppose, for
simplicity of explanation, that the values d and K be considered
unity in all cases. We thus have,

$$W = A \times V. \quad . \quad . \quad . \quad . \quad . \quad . \quad (2)$$

To solve this equation some one of the variables must be
constant and another determinable for a certain condition. If,
for example, the velocity "V" is assumed to have a constant or
fixed value, it will be seen that the value of " W " varies as the
area " A " or, conversely, for different quantities of flow the area
must increase or decrease if the velocity is to remain the same
under all conditions of flow. Now, before progressing further,
let us consider a very vital fact, namely, that the flow of fluid,
steam for example, through a pipe or orifice is produced by a
drop or lowering in pressure between the inlet and the outlet
or discharge. Just as hydraulics teaches that water seeks its

level, so equilibrium is established through increased velocity
in a pipe-line discharging steam.

The drop, therefore, is a function of the velocity; and with
a certain definite drop through an orifice of varying cross-
section the amount of steam passing through will vary with
the size of the orifice. The recognition of these two relations
led to the invention of a type of steam meters which have been
classified as "area meters," but which are also known by
various other names, such as:
displacement meters, cross-
section meters, throttling
meters, and valve meters.

The above simple mathe-
matical considerations upon
which are based the design
of area meters, will be better
understood by referring to
Fig. 25. This diagram comprises three elements: *A*, the
restricted orifice in a pipe-line;
B, the movable valve disc;
and *C*, the balance weight.

Suppose the system to be
at rest, i.e., no steam flowing.
The valve disc *B* would be
kept seated tightly by a
weight *C*. Now, if steam be
admitted to *A*, its pressure

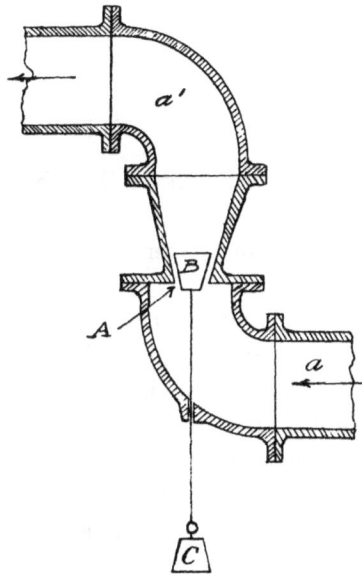

FIG. 25 —Diagram illustrating principle
of Area meters

distributed over the lower surface of the valve will tend to
overcome the weight *C*, permitting same to rise from the seat.
This allows steam to flow through the orifice and enter the chamber *a'*. Of course, the pressure of the steam will immediately
operate here on the upper surface of the valve and have a tendency
to counteract the upward rise of *B*. The area of these surfaces
being equal, it will be understood that the additional force *C*,
downward, will necessitate a greater pressure per square inch
at *a'* than at at *a*, if equilibrium is to be established.

According to the Bernouilli theorem, the sum of the pressure head, velocity head and friction head (or loss) must be constant in any two portions of a pipe containing a flowing fluid. If these factors are represented by $P, \dfrac{V^2}{2g}$ and F, respectively, then

$$P_1 + \frac{V^2}{2g} + F_1 = P_2 + \frac{V_2{}^2}{2g} + F_2 = K.$$

The friction loss through short orifices or pipes may be neglected and therefore we have the purely theoretical assumption:

$$P, \text{ (Potential Energy)} + \frac{V^2}{2g}, \text{ (Kinetic Energy)} = K.$$

from which it follows that the various energies in the fluid per unit of mass, are interchangeable and susceptible of transformation or conversion into other forms, their total remaining a constant.

This is what occurs in the example under consideration. At a' the protential energy (or pressure head) is lower than at a, the flow of steam through the orifice reducing same by a certain definite amount, depending upon the design of the different component parts, including the weight C. This drop in pressure $(Pa - Pa')$ is transformed into kinetic energy (or velocity head) and the result is an increase in velocity through the restricted area due to the throttling effect of the valve. Conversely, this throttling energy or pressure drop is utilized in lifting the valve from its seat. The valve is so tapered that the vertical movement of B is directly proportional to the increase in net area and, therefore, to the weight of steam flowing. This movement, when recorded by suitable devices, gives a continuous record of all momentary variations of flow. The theories stated above make it possible to deduce a workable formula for area or throttling meters.

For the factor V, Equation (1), we may substitute its equivalent as represented by the well-known form.

$$V = \sqrt{2gh} \quad \ldots \ldots \ldots \quad (3)$$

where
V = Velocity in feet per second.

g = Acceleration of gravity = 32.2 ft. per second.

h = Head in feet, of an imaginary column of the fluid, corresponding to the increase in velocity head, or which would be supported by its equivalent, the pressure drop.

The value h is determined, as we have just seen, by the pressure drop through the orifice, and may be computed as follows:

The pressure per square inch exerted by a column of fluid 1 ft. high and having a density d pounds per cubic foot is equal to $\dfrac{d}{144}$, therefore:

$$h = (p_1 - p_2) \div \frac{d}{144},$$

where
p_1 = Initial pressure—below the valve—pounds per square inch absolute.

p_2 = Discharge pressure—above the valve—pounds per square inch absolute.

Therefore equation (3) becomes,

$$V = \sqrt{2g\frac{(p_1 - p_2)}{\frac{d}{144}}}$$

$$= \sqrt{2g\frac{(p_1 - p_2)\ 144}{d}}, \quad \ldots \ldots \quad (4)$$

Substituting in (1)

$$W = A \times d \times K \times \sqrt{2g\frac{(p_1 - p_2)\ 144}{d}}$$

$$= A \times d \times K \times \sqrt{2g \times 144} \times \sqrt{\frac{p_1 - p_2}{d}}$$

Clearing this equation and combining constants,

$$W = A \times K \sqrt{d(p_1 - p_2)}$$

and since $(p_1 - p_2)$ is a constant for any one meter, depending only on the arbitrary weight of the valve the equation may be further simplified by incorporating this value in the constant K. Accordingly,

$$W = A \times K\sqrt{d}, \quad . \quad . \quad . \quad . \quad . \quad . \quad (5)$$

This formula represents the theoretical action of area meters and has been found to be in remarkable agreement with actual test results throughout practically the entire range of pressures for which the meter is designed. Thus, if we know the characteristics of each meter as it leaves the manufacturer; viz., the the drop in pressure $(p_1 - p_2)$, and the net area of the orifice for certain rises of the valve, the performance of the meter may be very closely approximated by calculation. However, there is no necessity of this in practice, since each meter is calibrated before leaving the factory, by passing steam through it and thence into a surface condenser, enabling measurement of the condensed steam to be made on suitable platform scales.

An empirical formula extensively employed by engineers for computing the flow of steam through an orifice is given in Babcock & Wilcox " Steam."

$$W = A \times K\sqrt{p_2(p_1 - p_2)}$$

in which the notations are as given in equations (4) and (5). This formula may be reduced to the form,

$$W = 1.7 \times A\sqrt{p_2(p_1 - p_2)}, \quad . \quad . \quad . \quad . \quad (6)$$

and will be found to check favorably with formula (5)

St. John Meter. This meter, first patented about 1893, is representative of the area type of meter and will be described in detail, as it has enjoyed perhaps more popularity among steam companies in this country than any other meter of this type. The same type is manufactured in Germany under the names of Baeyer, Rhenania, etc. and the following description will suffice for all.

Most engineers are familiar with the action of the pressure reducing valve in which the steam flowing in a pipe-line is throttled or wire drawn through an opening or orifice which is smaller than the size of the inlet pipe. The steam after passing through this restricted orifice expands to a lower pressure on the outlet side of the valve. This throttling action is controlled by means of a diaphragm with adjustable tension spring or weighted lever arm so as to maintain any desired throttling effect on the valve and reduce the discharge pressure to any extent, regardless of the initial pressure. This produces the " drop in pressure " referred to above. Equation (3).

Generally speaking, the action of the St. John meter and other meters of this type may be compared with that of the reducing valve just described. The meter is so constructed that the function of the springs and diaphragm is taken care of by the tapered valve or plug which is mounted on a vertical guide-shaft, free to move in a vertical direction in proper openings; and so proportioned and of such weight that, with a constant throttling effect or pressure drop of about 2 lbs. per square inch through the valve seat (which in this case is called the orifice) the valve or plug will be caused to float freely in the path of the flowing steam, its height from the valve seat being proportional to the quantity of steam flowing at any instant. Before proceeding further with this description let us refer to Figs. 26 and 27 which show the detailed construction of the meter.

Description of Meter. The meter casing is made of three castings as follows:

The meter body *B* to which are bolted the inlet casting *A* and the outlet casting *C*.

The other parts of the meter are as follows:

The brass valve-seat *S* which forms the perimeter of the orifice space and is screwed into the inlet casting *A*.

The meter valve *V* consisting of five parts:

(1) Tapered brass-plug having a diameter at top exactly fitting the seat *S* tapered down to the proper diameter at the bottom, by which the area of the orifice is increased in proportion to the vertical rise of the valve on its spindle or shaft.

(2) Valve disc bolted to plug (1).

(3) Valve spindle, or guide shaft;

(4) Dash pot-piston with cast iron snap-rings;

(5) Dash pot.   This dash pot must be at all times partially
filled with condensed water which provides a cushion to prevent
undue pounding of the valve in its seat.

FIG. 26.—St. John meter   FIG. 27.—Construction of St. John meter

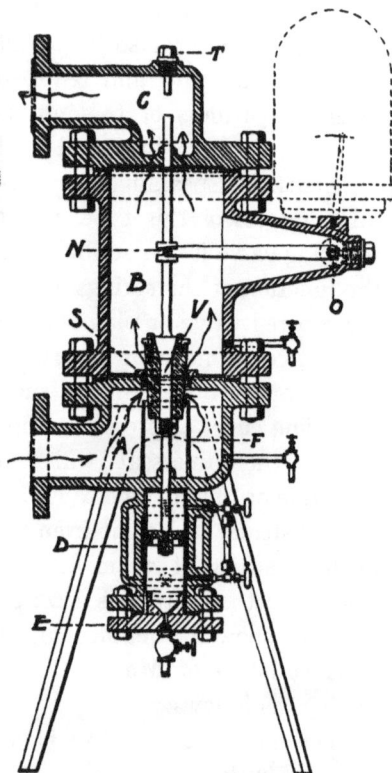

The bushing E or opening, in the inlet casting is to receive
and guide the vertical shaft, or piston-rod (4).

F is a deflector which is a part of the inlet casting and
serves to prevent the impinging of the steam directly against
the side of the tapered valve.

T is an adjustable screw by means of which the rise of the
valve in a vertical direction is gauged so as to allow about 2 or
$2\frac{1}{2}$ in. maximum vertical movement, thus determining the maximum capacity at which the meter can be operated.

N is a yoke on the shaft which engages a lever pivoted at O, by means of which the vertical movement of the spindle is communicated to the pencil-arm, which in turn traces a line on a paper chart revolved by means of clock work. The pencil-arm also has a pointer which moves across the face of the dial the latter being marked to indicate the horse-power of the steam flowing at any instant.

Operation of Meter. The meter is inserted at a convenient point in the steam line and the steam enters the lower inlet connection A and leaves at the top outlet connection C, which is connected again into the main steam line. When no steam is flowing, the tapered valve or plug V, will seat itself due to its own weight and remain tightly set against the valve seat S. While in this position, the yoke N and the lever arm O will be in their lowest positions, consequently the pencil-arm and the dial reading will record zero. When steam is admitted at A, the valve V will tend to rise after the pressure has built up sufficiently to overcome the weight of same. Because of the tapered form of this valve, as shown in Fig. 7, it is seen that the vertical movement of same will increase the orifice area from zero to the maximum when the valve spindle is raised to the valve stop T. The rate of increase of area depends solely upon the amount of taper given the valve and it is upon these factors that the quantity of steam is determined.

The weight of the valve V together with all its attachments, must be such that it may be exactly balanced and able to float and preserve its equilibrium in the path of the steam when there is exerted a differential pressure of about 2 lbs. per square inch between the inlet and outlet chambers. Thus is established the principle set forth above that if this difference of pressure is at all times maintained regardless of any position the valve may assume, the quantity of steam will vary as the net area of the orifice. Furthermore, it is evident that the orifice area is proportional to the rise of the valve, consequently this rise is a direct indication of the quantity of steam at any instant. This vertical rise, when transferred by means of the lever and pencil-arm will trace a line on a paper chart giving a continuous record of the flow of steam during any desired period not to exceed two weeks' duration.

Before this record can be reduced to a working basis, it is necessary to have a calibration constant. Therefore, each meter before shipment from the factory is calibrated by passing steam at various pressures through the meter, the outlet of the meter being connected to a surface condenser and the condensation weighed on platform scales. The constant so determined, when multiplied by the average height of the line drawn by the pencil-arm on the chart and the time in hours represented by the movement of the chart gives the number of pounds of steam which have passed through the meter. All meter constants are given by the manufacturer for a pressure of 100 lbs. gauge, and if the meter is to be operated at a higher or lower initial pressure, a further correction factor must be employed to obtain the number of pounds of steam which have passed through the meter. This factor is based on a curve which represents the square root of the density of the steam. A pressure recording-gauge must be used so that the average pressure during any stated period may be ascertained.

The recording chart is removed at regular intervals and delivered to the office of the heating company where it is measured by a planimeter, this being the most accurate means of determining the mean height of the pencil-line from the base-line, or zero-line, and being similar to the operation of measuring an indicator card from a steam engine. From this computation, the amount of steam the customer has used is ascertained.

These meters are made in capacities ranging from 27 to 1000 B.H.P. on the basis of 30 lbs. of steam per H.P. per hour. Where this meter is used on a non-fluctuating class of service, it has been found that after long years of service moving parts are in practically as good condition as when the meter was set. It is necessary to exercise considerable discretion when turning steam on the meter to prevent excessive vibration of the valve. This meter is very bulky and its use necessitates considerable changing of steam and pipe lines. The cost of the meter prohibits its use except for large users of steam. The maintenance cost is very small.

Sargent Steam Meter. This meter as shown in Figs. 28 and 29, is for the purpose of indicating the instantaneous flow of steam through the pipe-line in which it is inserted. The construction is very simple and consists of only two moving elements. The steam entering at a discharges through the circular openings b in the periphery of the tapered chamber c. Fitted over the outside of this chamber is an inverted cone-shaped bell g which seats at d when no steam is flowing. The

FIG. 28.—Sargent steam meter

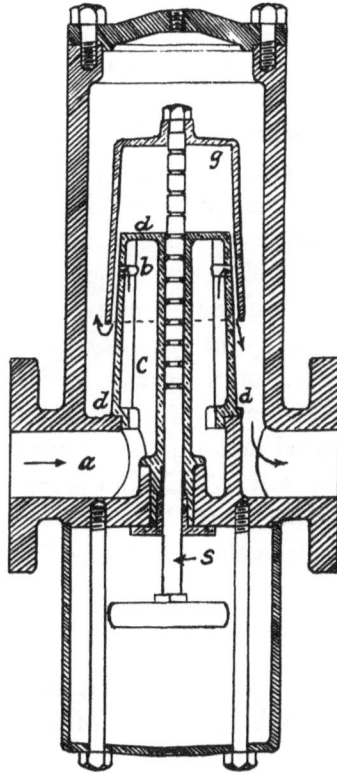

FIG. 29.—Construction of Sargent steam meter

steam flowing through the openings b forces the bell g upward and steam escapes through the orifice at d. The velocity being constant the flow is proportional to the rise of the bell from the seat. Since the net effective pressure surface on the under side of the bell is less (about 2%, due to the area taken up by the valve stem S) than that in the upper side, it will be seen that this constant pressure difference will govern the velocity of the steam. As we have just noted a constant pressure-drop produces a constant velocity, therefore the quantity of steam varies with the orifice area as in the St. John meter. The meter is equipped with a chart showing directly the quantity of steam flowing by means of an indicating pointer attached to the movable valve or bell stem. To compensate for changes in pressures and density of the steam, a device similar to the Bourdon pressure-gauge is used in such a manner as to govern the sidewise movement of the pointer. The chart is made up of curves following the density of corresponding pressures, therefore the pointer will intersect with the proper curve and the amount can be read directly without correction for pressure. Since this meter is simply an indicating device it is suitable only for testing purposes, no permanent record being made whereby the quantity passing through the meter in a given period of time may be integrated. It is therefore of little importance for commercial purposes. All meters must be calibrated and adjusted carefully before being used. See Trans. A.S.M.E., Vol. XXXI, p. 246.

Theory of Velocity Steam Meters. Referring again to the fundamental equation,

$$W = A \times d \times V \times K, \quad \ldots \quad \ldots \quad (1)$$

we find that a different solution is possible than that arrived at as the basis of Area Meters; namely, that we may assume a certain orifice of constant cross-sectional area A, and allow the changeable steam flow to be measured in terms of the resultant variation in velocity. This law forms the basis of the so-called *velocity meters*, as given in Table II. The basic principle of Velocity Meters is, in the last analysis not essentially at variance with the Area Meter, since there is involved in either the element of pressure differential. This type, however, embodies the application of the well-known idea of the pressure equivalent of velocity which is familiar to engineers generally, as exemplified by the Pitot-tube, Venturi-tube, Throttling-disc and modifications thereof.

Pitot-tube. The Pitot-tube was first used by the inventor Pitot, about 1837, for the measurement of water in pipes. Since then it has been applied to other liquids and gases, and finally to measuring steam flow. While there are numerous modifications of the original design the principle remains the same. To illustrate, in Fig. 30, let $A-B$ represent a pipe or conduit through which water flows with a velocity V feet per second. Insert two open tubes into the pipe as shown, one marked D having its opening pointed against the flow V; the other tube S, having its opening at right angles to the flow. The tube S is known

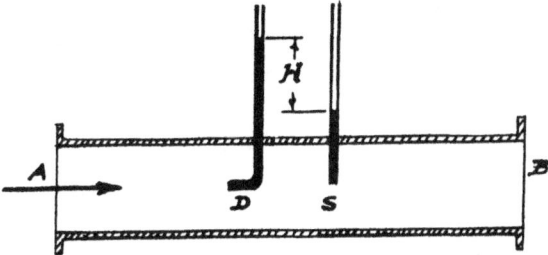

FIG. 30.—Diagram showing principle of Pitot-tube

as the static-tube and at all times is subjected only to the pressure due to the static condition. In other words, it is independent of the velocity (neglecting for the moment the slight aspiration or suction effect). The tube D, known as the dynamic-tube is also subjected to the static pressure and under no flow conditions this pressure will equal that of the static-tube. When a velocity is created, however, the dynamic-tube receives the impact of the moving particles of fluid in addition to the static pressure and thus water will rise in the tube D to a height H above the level of the static-tube, upon which, as we have just observed, the impact has no influence. The height H is an exact mathematical function of the velocity V, being the potential equivalent of the kinetic energy due to the velocity. The mathematical relation of these phenomena is represented by:

$$V = \sqrt{2gh}; \quad \text{or} \quad h = \frac{V^2}{2g}, \quad \ldots \ldots \quad (3)$$

where

V = Maximum velocity of flow in feet per second.

h = Difference in level in feet of water column.

g = Acceleration of gravity = 32.2 feet.

The foregoing will be understood as applying only to the flow of water. When it is attempted to measure velocity of steam and other gases by this method more complicated calculations are necessary. The reason for this lies in the substance used in the tubes to indicate the unbalanced pressures. Water is not a satisfactory agent for accomplishing this, and oils and mercury columns have been employed, the latter having proven the most successful. Thus the simple equation given must be elaborated so as to consider the different densities of the flowing and the indicating substances. If mercury is used instead of water, to indicate the velocity, the solution is as follows:

$$Hw = Hm \times \frac{dm}{dw}. \quad . \quad . \quad . \quad . \quad . \quad (7)$$

in which,

Hw = Height of water column.

Hm = Height of mercury column.

dw = Density of water.

dm = Density of mercury.

Substituting in (3):

$$V^2 = 2g \times \left(Hm \times \frac{dm}{dw} \right); \quad \text{or}$$

$$V = \sqrt{2g \times \left(Hm \times \frac{dm}{dw} \right)}, \quad . \quad . \quad . \quad . \quad (8)$$

$$\sqrt{2g \times \left(Hm \times \frac{13.6}{1} \right)}$$

$$= 29.6 \sqrt{Hm}, \text{ feet per second.}$$

This value when substituted in (1) gives the quantity of fluid flowing per unit of time. This formula, as will be seen later, applies to the various other types of velocity meters, including the Venturi. The above considerations are theoretical, to be sure, but actual practice agrees approximately with the results given. By substituting proper values for density, we may in a similar manner compute the flow of steam, using either water or mercury indicating columns since each is heavier than the measured substance.

Description of Pitot-tube Devices. A great many years of experimentation were required before any headway was made in the art of metering steam by variable or natural velocities. After Holly's first efforts, the first recorded attempt at improvement may be credited to Lindenheim, who about 1896 constructed a meter combining a modification of the Pitot-tube, as a primary, with a form of rotary similar to the Holly meter, Figs. 32 and 33. In reality it differed only in that a shunt was employed (see Fig. 34), thus obviating the necessity of

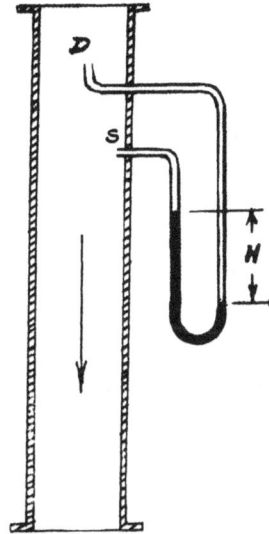

FIG. 31 —Mercury-manometer for measuring velocities

passing all the steam through the meter. These two types of meters were never really successful inasmuch as they were subject

FIG. 32.—Holly steam meter

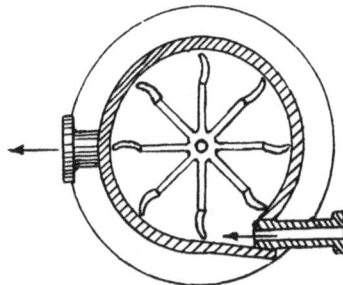

FIG 33.—Construction of Holly steam meter

to excessive wear, due to rapidly revolving parts. One serious drawback lay in the fact that they would not operate on pres-

sures much below 25 lbs. per square inch. Their accuracy may
be seriously questioned and they have largely been supplanted
by the improved forms which followed. Such devices are
generally accurate when operating at or near their tested and
rated capacities, but will under-register at low loads and spin
or over-register at overloads. These variations led to much
unfavorable criticism of the meter and led to their abandonment
where reliable results were imperative.

Some twelve years later, 1908, the Gebhardt Indicating
Pitot-tube meter was proposed. This meter differed in form
from Lindenheim's by the removal of the revolving element and

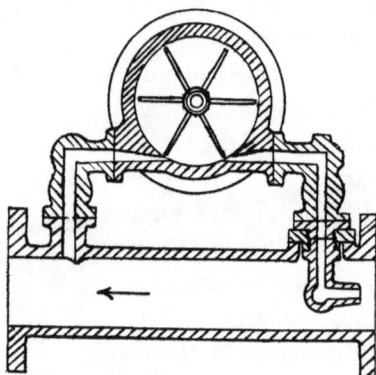

FIG. 34.—Lindenheim steam meter

substituting a moving part which is deflected by the impact
of the steam jet through a limited range of arc. The meter just
described—including the Holly and Lindenheim,—to quote
Prof. Gebhardt are, " In point of compactness, simplicity, ease
of application, and the integrating feature of the mechanism,
the ideal type of velocity meter, but it is probable that its
inherent limitations will prevent it from ever becoming a commercially successful instrument." This opinion has been amply
verified in recent years, as inventors have directed their efforts
along other lines, and no new devices of this type are being
manufactured.

Water Manometer. Unsuccessful attempts to introduce the
impeller type of meter stimulated activity in other lines leading

to the experiments of Burnham, 1906.* The earliest and most primitive form experimented upon consisted simply of a Pitot-tube leading into a common gauge glass, shown in Fig. 35. It is needless to say that these never attained importance until greatly modified and improved upon. By constant experimentation this meter has evolved from the crude design shown above to the form known as the Clyde or Gebhardt steam meter, which is now one of the very few meters employing water as the indicating substance. This meter is the most inexpensive meter to be obtained, but as its use is limited to the indicating feature exclusively, it is employed only for testing purposes.

Gebhardt Indicating Meter. This meter, shown in Figs. 36 and 37, solved the difficulties of the water-manometer type and perhaps it will be of interest to describe it in detail, as one of the few commercial devices now in common use.

The steam when first introduced through the dynamic tube into the dynamic chamber at the bottom condenses, forming

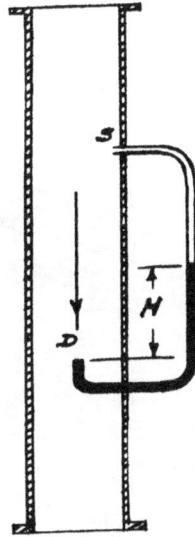

FIG. 35.
Water-manometer

water entirely out to the tips of the tube, which is provided with a number of small openings at right angles to the flow; both tubes being built together, and inserted through the wall of a steam-pipe in a very simple manner. When no steam is flowing in the pipe, the condensation will rise only to the height of the dynamic tube, giving a zero reading, but when steam is flowing the column of water rises as a result of the unbalanced differential, as has been explained before. If we measure the height of the water column, at any instant, we have again approximately,

$$\frac{V'^2}{2g} = Hw.$$

* Armour Institute of Technology.

FIG. 36.—Gebhardt indicating steam meter

FIG. 37.—Detail of Gebhardt indicating steam meter

and since we are now dealing with steam as a flowing substance, instead of water as in equation (3) we have.

$$\frac{V^2}{2g} = Hw \times \frac{dw}{ds}, \quad \ldots \quad \ldots \quad (9)$$

where

dw = density of water in pounds per cubic foot.

ds = density of steam in pounds per cubic foot.

The solution of (9) results as follows:

$$V = \sqrt{2g \times Hw \times \frac{dw}{ds}}, \quad \ldots \quad \ldots \quad (10)$$

$$= \sqrt{2g \times dw \times \frac{Hw}{ds}}.$$

$$= \sqrt{2 \times 32.2 \times 59.76 \times \left(\frac{Hw}{ds}\right)}.$$

$$= 62.04 \sqrt{\frac{Hw}{ds}} \text{ feet per second.} \quad \ldots \quad (11)$$

Equation (11) represents the maximum velocity of the flowing steam. It is well known that the velocity diminishes near the edge or inside surface of a pipe due to friction, therefore, the point of mean average velocity for the section will be found somewhere between the maximum (at the center) and the wall of the pipe. This factor, which will be called K must be determined by experiment for each size pipe. The mean velocity then becomes,

$$V = 62.04 \times K \sqrt{\frac{Hw}{ds}} \text{ feet per second.} \quad \ldots \quad (12)$$

After determining the velocity as above the value may be substituted in the formula,

$$W = A \times d \times V \times K \quad \ldots \quad \ldots \quad (1)$$

Then,

$$W = A \times ds \times 62.04 \times K \sqrt{\frac{Hw}{ds}} \text{ lbs. per sec.}$$

$$= A \times 3600 \times 62.04 \times K \sqrt{(Hw. ds)} \text{ lbs. per hr.}$$

$$= 223344 \times A \times K \sqrt{(Hw \times ds)}. \quad \ldots \quad (13)$$

Formula (13) is the equation used for computing the charts, furnished with the meter. These charts, as will be seen, are mounted upon a cylinder placed at the side of the water glass, and are graduated to read in pounds of steam per hour directly.

The Republic Steam Flow Meter. This meter, Fig. 38, is the outgrowth of the Gebhardt Indicating water-column meter. The first attempt at improvement—with the idea of making the meter a recording device—consisted in the application of the water column of a Pitot-tube in contact with a resistance coil. The vertical movement of the water column, proportional to the velocity of steam causes a varying amount of current to flow through the coil and registers through a recording ammeter and integrating watt-meter. The readings from these instruments when multiplied by proper experimental constants for different pressures give the quantity of steam flowing. Unfortunately, difficulty was encountered with the water column, it being found that impurities in the steam,—boiler compound and oils,—coated the coils and made correct results impossible. Finally, mercury was adopted with much better success, and by interposing an oil seal between the electric wires and the tubes, no water is permitted to interfere. (These instruments have not been on the market long enough to have proven their worth, but it appears they will be acceptable for commercial-use.)

Fig. 39 requires very little explanation. Each conductor in the set has a specific resistance which regulates the flow of current so as to register according to the square root of the height of the mercury column. The current for actuating the instruments is supplied by a 30 volt motor-generator connected to a lighting circuit. The secondary element may be placed at any desired distance from the point where the steam is metered, and as electrical instruments are very accurate this method will appeal to many engineers, notwithstanding its seeming delicacy and complication.

The General Electric Steam Flow Meter. In 1910, some years prior to the introduction of the Republic meter, another form of Pitot-tube meter was brought out by the General Electric Company. After experimenting on an extensive scale, this

FIG. 38.—Republic steam-flow meter

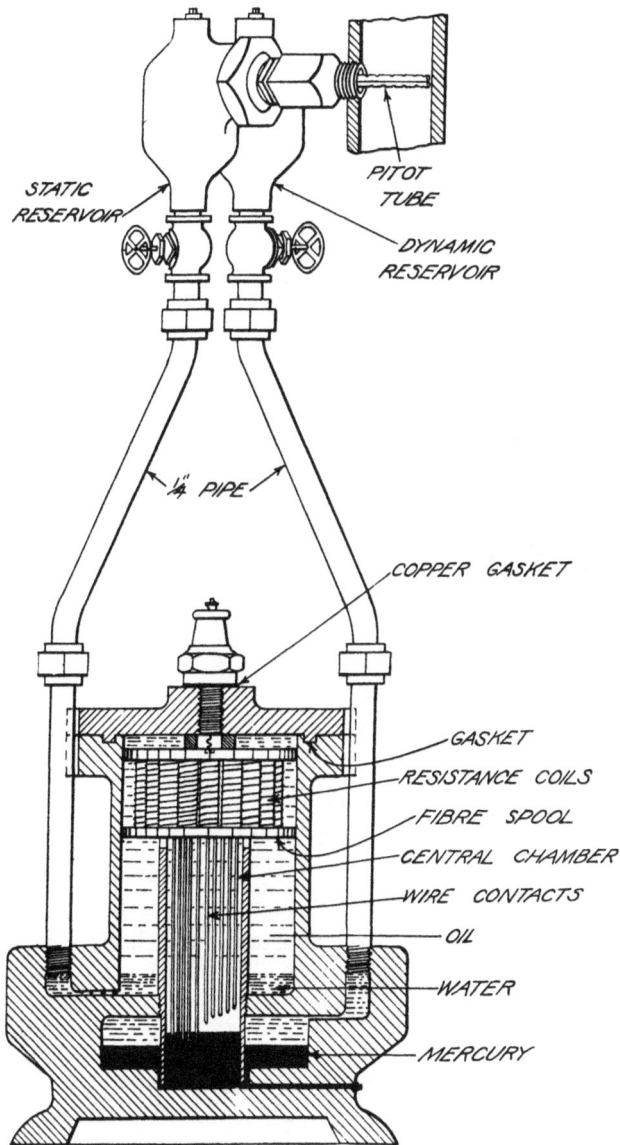

STATIC RESERVOIR

PITOT TUBE

DYNAMIC RESERVOIR

¼" PIPE

COPPER GASKET

GASKET

RESISTANCE COILS

FIBRE SPOOL

CENTRAL CHAMBER

WIRE CONTACTS

OIL

WATER

MERCURY

Fig. 39.—Construction of Republic steam-flow meter

company has perfected a velocity meter embodying many admirable features. There are many of these meters in operation, principally for large steam consumers. Considering this fact, it will probably not be amiss to enter into a detailed description of the device.

The primary element, known as the " nozzle-plug " resembles substantially that employed by the Gebhardt and Republic meters, just described. The flowing fluid impinges against the leading opening, Fig. 40, setting up a pressure equal to the static pressure plus that resulting from the dynamic or velocity head. A suction effect is exerted on the trailing opening, the net result being a differential pressure which reaches the mercury manom-

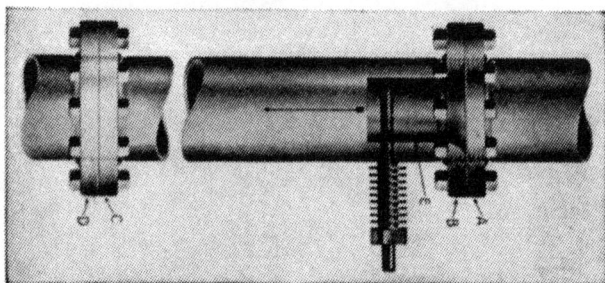

FIG. 40.—Pipe reducer for increasing steam velocity

eter. These nozzles, it must be understood possess certain peculiar characteristics which differentiate them from other forms of Pitot-tubes used in other meters and therefore in calibration we would expect the design of these tubes to have an important bearing upon the values of the constants used in the theoretical formula. In the elementary form of Pitot-tube the single dynamic opening is located at the center of the pipe, and also as a result, gives the maximum velocity. In the modified forms we are considering the openings are spaced entirely across the pipe and thus the velocity indicated represents more nearly the true average.

For measuring the flow in pipes less than 2 in. in diameter, orifice tubes similar to the Venturi form are inserted in the line and the differential pressures communicated to the recording mechanism as will be described under " Venturi," p. 143.

When the velocity is very low, as is the case where the piping is much larger than necessary, a pipe reducer, or brass

FIG. 41.—General Electric steam-flow meter

funnel, as shown in Fig. 40, may be used to excellent advantage. This serves to increase the velocity with consequent higher differentials, and also is an insurance against variations in the diameter of pipe, which will seriously affect the accuracy of the meter.

Secondary element. The body of the meter is constructed of iron, designed to form a manometer arrangement Figs. 41 and 42. Upon the surface of the mercury chamber N, a float O, rests. The stem of this float engages a pinion Q which is on a shaft attached to magnet S. A similar magnet H is mounted

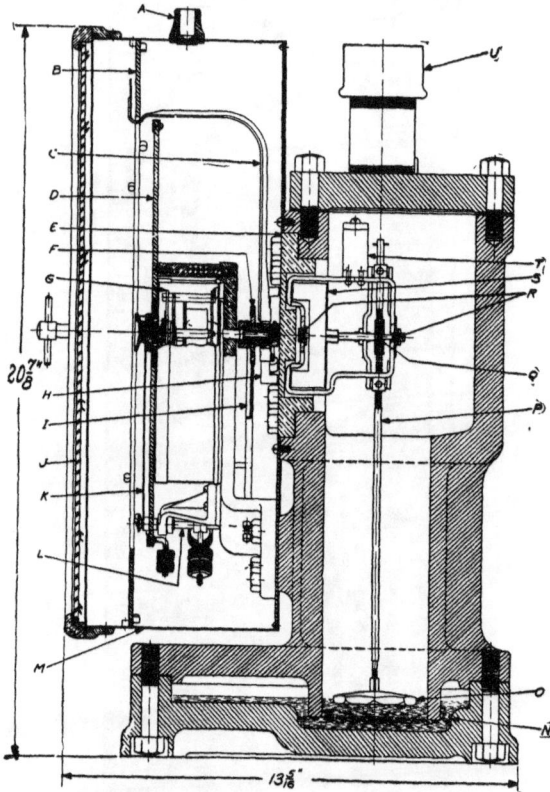

Fig. 42.—Construction of General Electric steam-flow meter

on another shaft on the side opposite the mercury float chamber. Any vertical movement of the mercury occasioned by the differential pressure will cause a rotation of the two permanent magnets, eliminating the friction which would be unavoidable if a packed stem were used. The second or outside magnet actuates the indicating pointer C, and pen point K, which records on the chart mounted on disc D, and revolved by clockwork G.

The indicating scale is calibrated to read in any desired unit. The scale of divisions on the chart is arbitrary and must be multiplied by a constant after the average ordinate of the curve is determined. This constant must be determined by the diameter of pipe and steam pressure.

In the earlier forms of the G.E. meter, a device was supplied for automatically correcting for pressure variation but in the latter types this feature has been omitted. A form of planimeter attachment is furnished for automatically integrating the flow of steam as recorded by the pen on the chart. The rim of a friction wheel rests against the surface of the paper chart and the speed of rotation is determined by the variable angle of its plane from tangent to radial positions as considered with reference to the center of the chart. This rotation is recorded on three small dials, similar to a revolution counter and the successive readings are multiplied by a constant. Unlike some other meters, the radial ordinates on the chart are not uniform but vary approximately as the square of the velocity, thus a special radial planimeter is furnished to average the true value of the ordinates. The integrater, however, through a cam arrangement is direct reading.

The Hallwachs Meter. Another modification of the Pitot-tube is to be found in the Throttling-disc type which is extensively used in Germany where it was first introduced. As yet very few meters of this type have been used in this country. By referring to Fig. 43, it will be observed that a disc with a bore or orifice slightly smaller than the inside diameter of the pipe is inserted between a pair of flanges by shortening or springing the piping. This has the effect of dividing the pipe into two vessels and the flow of steam through the orifice will be found to agree with the formula given by Zeuner. Equation (6) is a modified form of this same equation. The drop in pressure varies of course with the velocity and the differential actuates a mercury column. As the mercury rises it comes into contact with platinum wires fused into the glass manometer tube. These wires connect to resistance coils of carefully computed value and adjusted so that the rise of the mercury, although

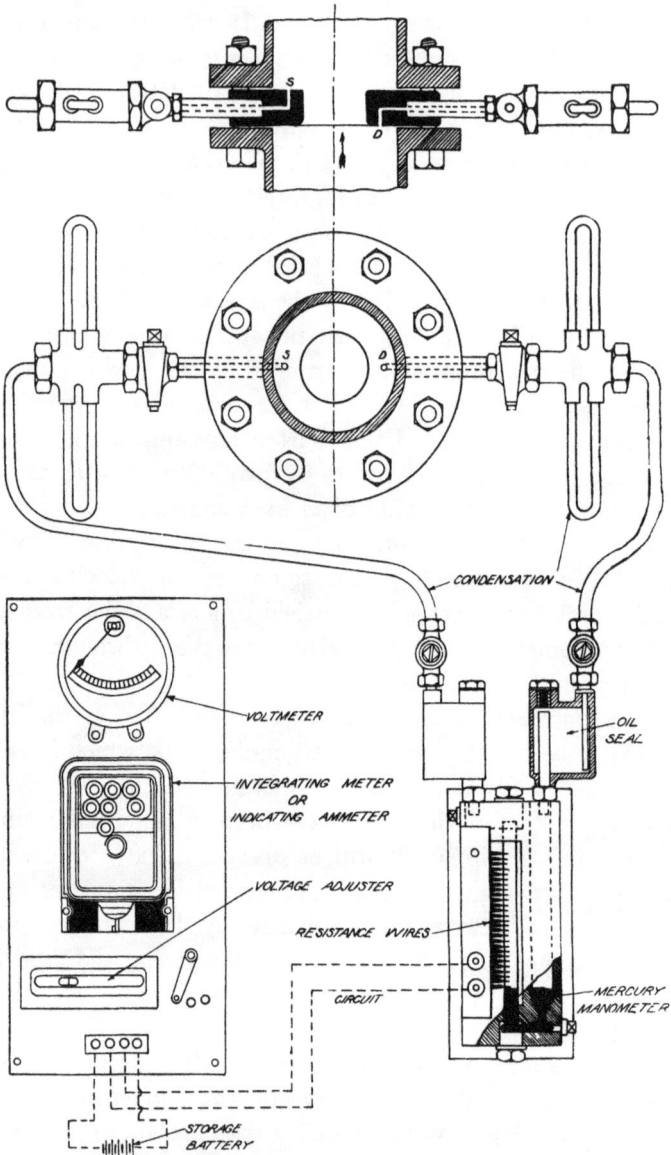

FIG. 43.—Construction of Hallwachs steam meter

following the square root law, or approximately so, will cause a direct registration on the electrical recording instruments. The secondary element of this meter is similar to that of the Republic type. It consists of a recording instrument of high accuracy and an ampere hour meter for integration. The meter is rather delicate as judged by standards in this country, is relatively high in cost on account of import duty and is not advised for other than testing work requiring exceptional accuracy for these reasons. The electrical current is supplied by a storage battery.

As indicated in Table II, there are other forms of instruments made to operate on the Pitot-tube principle but these embody no new principles, with the exception of the Bristol meter. In this device the special shape of the mercury well compensates for the square of the velocity giving a direct rise to a float which records on an instrument similar to the ordinary recording steam pressure gauge. This instrument is as yet in the experimental stage.

Before leaving the subject of velocity meters, it will be well to consider the Venturi tube, and its application to the measurement of water and steam.

Venturi Tube. The phenomena described below was first observed by the Frenchman, J. B. Venturi, in 1791. He did not recognize the utility of the principle he had discovered and made no use of it. It remained for Clemens Herschel, almost a century later (1887), to apply this knowledge in a practical way. The result is the Venturi meter which has been used very extensively since that time for measuring water. The meter has since been modified for the purpose of metering other liquids and gases, including steam. Fig. 44 illustrates the principles governing the action of the primary element or tube, and Fig. 22 shows the secondary element attached.

A fluid flowing through a Venturi tube must pass the contraction at B at a greater velocity than at A. After passing the point B, the velocity due to the suction effect of the expanding nozzle, returns to the normal with only a perceptible loss in pressure at C. The increased velocity at B results in a lowering of the pressure at that point, and if a differential manometer is connected between A and B, the difference in the heights H, of the two columns will be a measurement of the velocity of the fluid. This height will increase approximately as the square of the throat velocity, $V^2 = 2gH$; if the velocity is doubled therefore, the height will be four times as great. Therefore,

$$W = K \times A \sqrt{2gh}. \quad . \quad . \quad . \quad . \quad . \quad (14)$$

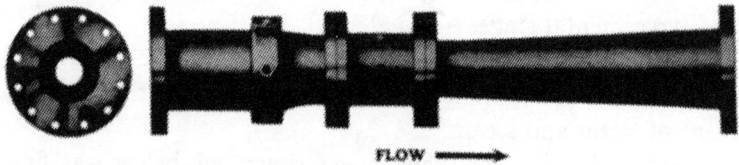

FIG. 44.—Venturi Tube

Where

$W =$ Quantity flowing per second.

$K =$ Experimental constant which has been found to be within 2% above or below 0.98 for water at average velocity.

$A =$ Area of throat of tube in square feet.

$H =$ Head in feet corresponding to the difference in the pressure of the fluid entering the tube and the throat.

The Bernouilli theorem applies also in the case of the Venturi tube. A portion of the static pressure head at A is converted temporarily into dynamic velocity head through the throat B. The function of the downstream (discharge) nozzle is to regain the changed energy of the pressure head through a lowering of velocity. This is done with only slight permanent loss of pressure.

It is evident that the high ratio of contraction in area at the throat will produce relatively large differentials between B and C. This renders it necessary to use a mercury manometer, and with extremely high velocities the difference in level produced when measuring water will be as high as 22 in. corresponding to a differential of

$$22 \times 0.4908 = 11.8 \text{ lbs. per square inch.}$$

Thus it will be seen that the facility of accurately reading the manometer levels or of automatically communicating same to recording devices is greatly enhanced by the great range of measurement as compared with the much lower differentials obtained by the Pitot-tube.

The coefficient of this instrument at average load is 98%. At low velocities the coefficient is greater and at high velocities lower, varying through the guaranteed range by not more than 2% above or below. The Venturi water-meter is extensively used for boiler feed water service, and if care is observed in selecting a tube of proper proportions, it gives very satisfactory results. The accuracy is questionable below 8% of the maximum rate of flow, as the meter will under-register. As a steam meter, it is to be recommended principally for superheated steam, as moisture will affect the accuracy to a great extent.

III. CONDENSATION METERS

THE idea of measuring the condensation from heating systems to determine the steam consumption was a very radical departure from previous practice, and did not take concrete form until 1904 in which year a patent was granted to John D. Walsh on the device which has since been manufactured under the trade name—Simplex condensation meter. The introduction of this meter marked an important step in the evolution of the industry and while it cannot be said to have been a cureall for the weaknesses apparent in the operation of district heating-systems, yet the decade intervening since the idea first took root has furnished ample justification for the experiment.

The advent of the condensation meter ushered in an era of prosperity for many heating companies that had been struggling along in hopeless financial difficulties, not because of improper installations and management, but as a result of the inability to sell their product scientifically and secure adequate returns on the investment.

The first manufacturer to enter the field with a condensation meter was the American District Steam Company. Through its affiliations with extensive heating interests throughout the country this company, in the capacity of engineers, operators and owners has been naturally equipped for taking the leadership when improvements of any kind are demanded. It is not at all strange that their experience prompted them to experiment on some device which would be more satisfactory as between company and consumer than was the primitive Holly steam-meter; and at the same time offer an avenue of escape from the evils lurking in the various systems of flat rates.

Since it has been seen that steam meters are usually complicated, it follows that they will also be comparatively costly. It is obvious that for large companies serving hundreds of customers, the investment charges for steam meters would be enormous. Not only the first cost but also the amount of attention required and the labor of computing the meter records are factors entering into the question. For the larger customers, this factor is of no serious import, inasmuch as steam meters are available in capacities as large as desired, and the maintenance and office expenses are thereby reduced to a negligible amount in proportion to the total steam sold. For the multitude of small individual consumers, e.g., residences, stores and restaurants, the steam meter must be waived in favor of a much simpler method of measurement. The condensation meter has played a very important part in the steam-heating industry, much more so in fact than the steam meter, primarily on account of the difference in cost. The condensation meter fills the requirements for a simple, accurate and inexpensive method of selling steam and consequently has been used almost exclusively by most companies, except in the largest cities.

TABLE III. CLASSIFICATION OF CONDENSATION METERS

	Primary Element	Secondary Element	Name and Maker
Weight meters	Dump bucket	Gears and dials	Simplex. 1904. American District Steam Co. Crescent. W. H. Pearce & Co.
	Revolving drum	Gear and dials	Tyler. 1914. Tyler Underground heating systems.
Volume meters	Revolving drum	Gear and dial	Detroit. 1906. Central Station Steam Co.
Trap meters	Tilt trap	Trip-counter	Cranetilt. Crane Co.
	Float trap	Diaphragm-counter and dial	Trapand. 1912. J. C. Hornung.

Most of the above meters are acceptable for the purpose intended provided every precaution is taken to install them properly. The most essential point to consider is the fact that condensation meters are not guaranteed to operate under pressure; they are essentially gravity meters in that they will not discharge when there is any back pressure on the outlet or discharge line. Therefore, their use is limited to places where the following can be observed:

" (1) The returns from the radiation must pass through a steam trap of the constant or continuous-flow type, to prevent sudden surges of the return water. This liability is present when pot or bucket traps are used.

" (2) Some method must be adopted for relieving the discharge line from the trap of any vapor, which might otherwise enter the meter. Hot water vapor or air, in mixture with the condensation has a tendency to " spin " the revolving type meter; and, due to its pressure produces a velocity of impact on the dumping type. Both of these effects lead to inaccuracy, usually over-registering. Methods of overcoming this are shown in Figs. 45 and 46.

" (3) The meter must have a free and unobstructed discharge to a suitable catch basin if the returns are to be wasted to the sewer, or connected to a return line, which must be under gravity or vacuum pressure."

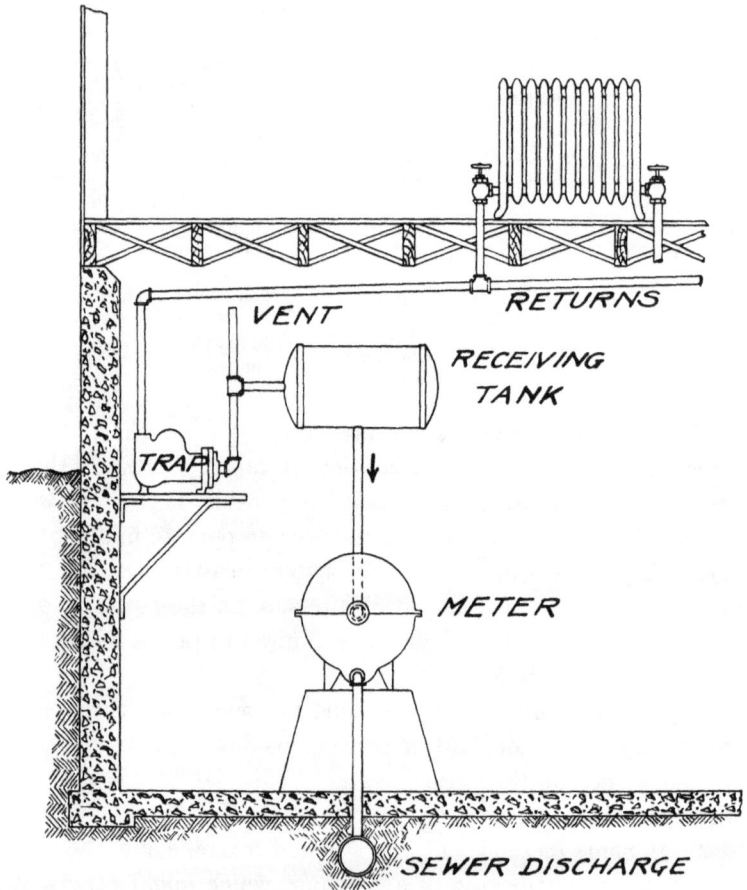

VENT

RETURNS

RECEIVING
TANK

TRAP

METER

SEWER DISCHARGE

FIG. 45.—Method of connecting-condensation meters

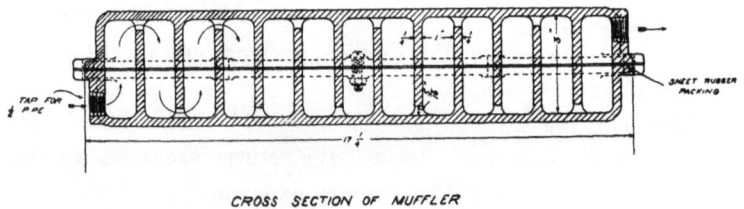

TAP FOR
½ PIPE

SHEET RUBBER
PACKING

17 ¼

CROSS SECTION OF MUFFLER

FIG. 46.—Device for preventing surge of condensation into meter

Simplex Condensation Meter. As shown in Figs. 47 and 48, this meter is of the " dumping type," having two buckets balanced on a shaft which rocks with them as they alternately fill and fall to empty the contents. The condensation entering from the inlet pipe at the top of the meter is deflected by a horizontal baffle plate so that it runs in a divided stream with lessened velocity into the receiving chamber. A nozzle with a restricted opening permits the condensation to flow in a thin stream to the main chamber of the meter, where a partition between the buckets turns the stream into the bucket which is to be filled. The added weight of the water causes this bucket

FIG. 47.—Simplex-condensation meter

FIG. 48.—Construction of Simplex-condensation meter

to overbalance the other and fall, and this position allows the water to drain from the lower bucket into the bottom of the meter, and thence through the outlet at either end of same to the sewer.

To the bottom of these buckets plungers are so fixed that connection is made with the dash pots in the bottom of the meter. These dash pots are filled with water at all times and a cushion is thereby formed which makes the meter practically noiseless in operation. The distance travelled by the bucket in its fall is determined by the adjustment of a screw upon the head of which the bucket strikes. The adjustment of this screw controls the capacity of the bucket. The oscillating motion from these buckets is transmitted to a revolution counter which records the condensation passing in pounds.

The bearings of the meter consist of bronze cups and cones which are protected by heavy felt washers and a cap at each end of the shaft supporting the meter buckets. Bronze balls are used in these bearings. Access to the bearings for adjustment may be had by the removal of the bearing cup on the front of the meter.

This meter will be found to be reasonably accurate between temperatures of 55° F. and 180° F. Generally speaking the average error of this meter will be slight, unless the meter is run considerably over or under the rated capacity. When the meter is operated considerably above its capacity the results are very uncertain, as the water runs through the meter too rapidly for accurate weighing. When the meter is operated below capacity the readings are less accurate, which may be due to the fact that at a very slow rate the water dripping slowly from the nozzle into the bucket has but little tendency to dump the bucket, as compared with the impact or velocity of a much larger volume of water passing the nozzle and striking the bucket at the same time, as is the case when the meter is operated above capacity. If the meter is carefully selected with regard to proper size and capacity and if it is properly installed and cared for, it should be sufficiently accurate and serviceable for the purposes for which it is intended, and will bear out the guaranty of the manufacturers fully.

Detroit Condensation Meters. These meters were first put on the market about 1906, and since then many of the defects present in the original designs have been eliminated successfully. The patent covering the application of the revolving drum or turbine to the measuring of fluids was issued to Hans Reisert, Cologne, Germany, and the rights were acquired by the present manufacturers in 1907, who have added several improvements to the original instrument. One of these consists in adapting the meter to vacuum return lines, which is accomplished by making an air tight drum casing. A suitable vacuum-pot or trap must be connected ahead of the meter to prevent leakage of steam into same. This feature does away with one of the principal difficulties in the way of a general adoption of condensation meters, viz.: adaptability to vacuum systems of heating, thus obviating the necessity of wasting the returns.

The mechanism of this meter, as will be seen by Figs. 49, 50, and 51 consists of a rotating cylindrical drum divided into six scroll-shaped compartments which receive one by one, in

FIG. 49.—Detroit condensation-meter

the order of their succession during the revolution of the drum, the flowing liquid as it is discharged from the inlet nozzle, or spout, which is introduced axially into one end of the drum and extends horizontally across but is not connected with it. The drum is made to revolve around this inlet nozzle as impelled by the weights of the liquid (water) in the different compartments. The inlet nozzle discharges this fluid from a long narrow opening along its bottom. Each compartment has an outlet at the perimeter of the drum for discharge of its contents at the proper time in the revolution. The compartments are filled successively and when completely filled, overflow into the next succeeding compartment. As each compartment acquires its weight of liquid, that weight, due to the design of the scroll-shaped compartments, shifts the center of gravity of the mass of water to one side of the perpendicular to the axis of the drum, tending to revolve the drum and empty the preceding compartment while the succeeding compartment

Fig 50.—Detroit condensation-meter with cover removed

takes its place directly under the discharge from the inlet nozzle and is in turn filled to overflowing and passed on and emptied by its successor. When compartment No. 1 is filled, the liquid overflows into compartment No. 2, which is next in rotation.

When accurately constructed, each compartment should cut off and overflow in exactly the same way, therefore, the amount of liquid passed should be exactly the same for each revolution of the drum and it is impossible for any liquid to be discharged without the rotation of the drum. Each drum is calibrated in pounds of liquid per revolution, and a direct reading counter indicates the number of pounds passing through the meter. The drum has no moving parts such as valves, or adjustable apparatus. It revolves on hubs, or drum supports, which rotate on bearings fastened securely to the meter case. The bearings consist of a hanger supporting bronze rollers on pins. The front bearing has round-faced rollers to engage in the groove in the front hub. The rear bearing has square-faced rollers and the hub is not grooved. The round-faced rollers and the hub engaging each other, keep the drum in position, and the flat-faced roller will permit of expansion. The only moving parts

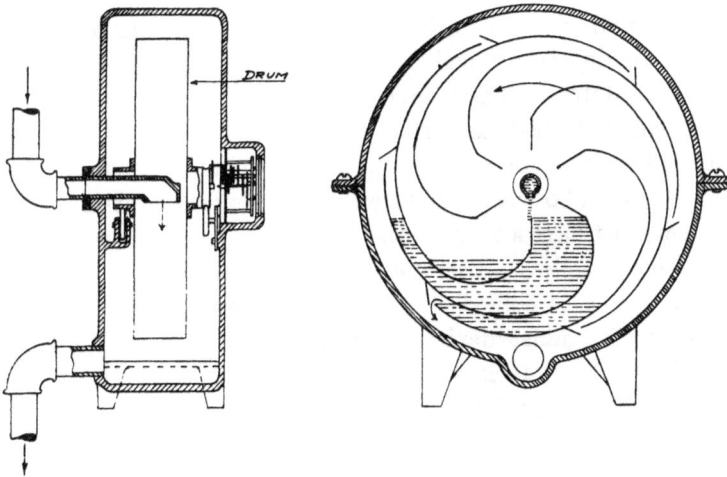

FIG. 51.—Cross-section of Detroit meter

of the equipment are the roller bearings, the gears of the revolution counter, and the drum.

This meter is customarily calibrated with water at a temperature of 140° F., and has proven to be accurate within 1%, which is certainly sufficient for all practical purposes, and bears out fully the guaranty made by the manufacturer. For wide ranges of load conditions the revolving drum is a very accurate type of condensation meter since within a certain critical limit, the entire volume of water must pass through the various compartments before reaching the discharge chamber. If an intelligent supervision is maintained over the installation, these meters will be found to give excellent results and will sustain their accuracy indefinitely.

A dependable feature of this meter is that the range of error is remarkably constant, or in other words, the error is fairly independent of the rate of the flow of water, even much beyond the prescribed limits of capacity. The motion of the drum is steady and will not continue after the liquid ceases to pass into it, if properly baffled from steam leakage. The bottom of the casing is allowed to remain partially filled with water which exercises a damping effect and overcomes partially the tendency of the drum to spin. In addition to this, baffles both external and internal with which the latest type of meters are equipped, practically guarantee against this evil.

Trap Meters. These devices are of relatively small importance commercially due partially to their high cost and the necessity of repeated calibration tests, also to the fact that they have never been actively introduced on the market. Since each system must have a steam trap ahead of the meter, these devices serve more or less satisfactorily the double function of a trap and a meter combined into one. The principal objection to be entered against devices of this character is the inherent liability of the valves and other moving parts of the trap to lose their accuracy. The advantage lies in their compactness.

LIMITATIONS OF ACCURACY OF THE DIFFERENT TYPES OF METERS

Water Meters. The displacement water meters are not recommended for sustained accuracy and repeated calibration tests are necessary. This is due to the fact that the moving parts are subjected to considerable friction wear, and in the case of hot water meters, the unequal expansion due to temperature variations causes the meter to stick under expansion, and to admit of slippage or leakage with lower temperatures under contraction. For use in short tests, these meters must be calibrated prior to and at the conclusion of the trial, the calibration conditions to approximate as nearly as possible the actual pressure and rates of running which are to prevail during the boiler test. All types of meters must be corrected for temperature variation.

The accuracy of velocity water meters is affected by the same factors and in a similar manner as is set forth under "Accuracy of Velocity Steam Meters," p. 157.

Weir meters which measure the volume of water are generally considered by engineers to be acceptable from the standpoint of accuracy. They are, in common with other types, limited on account of temperature variation. Some manufacturers claim to have overcome this difficulty and it is customary for them to guarantee accuracy within 1%. These meters are quite simple, the only moving element being the float submerged in the flowing water. These devices must operate under gravity conditions free from any air or vapor pressures, and are usually inserted in the suction of boiler feed lines, discharging into gravity chambers or hot wells before entering the pump.*

Automatic weighers are also guaranteed to be accurate within 1% when used under certain specified temperatures. A correction factor is necessary for variations in this respect. This type is used in a manner similar to the weir meters just described.

Steam Meters. Area steam meters, particularly the St. John have been found to give satisfactory results from the commercial heating standpoint. It would be impossible to guarantee any specified per cent of accuracy for this meter where it is used for measuring the supply of steam for pumps, engines, steam-hammers, etc. Furthermore, it is not adapted for this service, unless the installation is designed with extreme care. If the engine is at a great distance from the meter, these difficulties will be somewhat reduced but with all slow speed cut-off engines and with pumps taking steam full-stroke, the fluctuations are too violent to be measured accurately. In the case just mentioned the long discharge line from the meter would act as receiver or storage chamber and would minimize the effect of cut-off to some extent. These remarks do not apply with equal force to high-speed engines, since the

* See Trans. A. S. M E , Vol. 34, page 1055.

flow approximates more nearly a continuous stream. Central
Station companies seldom supply steam for such units, however, and usually the steam is used for slow-speed elevator
pumps, or similar apparatus. Extreme care must be used in
selecting a meter for this purpose, and in the computations of
the record charts, certain allowances are required.

Where the valve (area meter) must rise and fall rapidly in
response to the fluctuation of steam flow, it will be subject to
the following conditions, which impair its accuracy seriously:

"1. The effect of hammering on the seat which gradually
wears down the working parts, thus increasing the free area,
causing the meter to under-register, and

"2. The forces operating the valve or disc are being changed
from the ideal calibrated condition, and are affected by the
increased friction of the dash pot (see St. John meter) as well
as in overcoming the inertia of the floating valve. The St. John
meter is not recommended for use where the load averages less
than 10% of the normal capacity as it is liable to give inaccurate
results below this point."

It is advisable in some cases to install two meters and if the
load varies from day to night or from winter to summer, one or
two meters can be cut in as demanded. In order to secure
accurate results, the meter must be selected as nearly as possible
with just sufficient capacity for the amount of work it is to
perform.

Figs. 52 and 53 illustrate the forms of charts recorded on
different classes of service. These charts are frequently of much

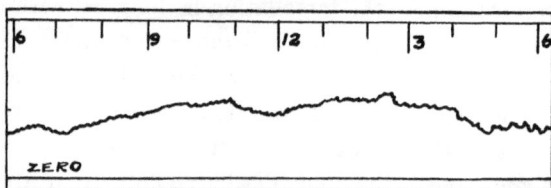

Fig 52.—Typical chart from St. John meter—steady load

value in adjusting misunderstandings which may arise between
the Company and Consumer. In large companies regular

inspection departments are maintained to care for these meters. The meter must be by-passed or shut off the line for a short period of time occasionally to determine if the recorder will

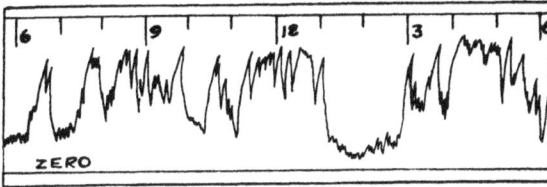

FIG. 53.—Typical chart from St. John meter—fluctuating load

touch the zero line. If this is not the case, it indicates some trouble with the working parts, which may become loosened after long service, or some foreign substance may be holding the valve off its seat permitting leakage of steam.

Velocity Meters. The inherent weakness of the velocity principle is the impossibility of securing consistent results throughout the entire range of operation. When the velocity is very low the manometer element is not sensitive enough to record the minute differential pressure and naturally the true velocity is unknown. The only solution, and one which is seldom applied because of the additional trouble and cost, is to employ different capacity tubes as in the Venturi, or smaller pipes for the Pitot-tube meters, these to be connected as a by-pass and used only when the load decreases. It is essential that the velocity attain measurable proportions at all times, as otherwise error is unavoidable.

It may be said that where the velocity falls to less than 20% of the rated capacity of the meter for the Pitot-tube type and below 10% for the Venturi type, the results will be inaccurate because of the extremely low differential operating on the manometer. Above this velocity meters are very successful if the flow does not fluctuate too violently. To partially overcome this fluctuation it is necessary to design special air chambers on the circulating pumps.

When mercury was first used as the indicating substance in manometers great difficulty was experienced because of its great density as compared with steam and water. To create any perceptible rise in the mercury column required very high steam velocities where Pitot-tubes were used. With the advent of the Hallwachs meter, using the throttling-disc, Fig. 43, a much greater difference in the heights of the mercury was possible because of the increased pressure drop through the orifice. Some of the later refinements in meters rely on a large pressure area on the surface of the mercury to force a much smaller column upward through a restricted area. By this method, a somewhat greater range is available, thus eliminating some of the liability of error in reading. The mercury column is less sensitive than the water column due to its greater density, 13.6 times that of water, and thus it will be seen that extreme accuracy is improbable at low velocities.

Where velocity meters are used very frequently an error is made in calculating the autographic chart. Since in the velocity type meter the differential pressures vary as the square of the velocity and as the amount of the steam used varies directly as the velocity, it will be necessary to extract the square root of the ordinates indicated on the chart and then multiply by the constants to secure the correct reading.

CARE AND MAINTENANCE OF METERS

If a steam heating business is to be conducted successfully, every reasonable effort must be made to preserve cordial relations between company and consumer; and the surest method of accomplishing this is to show a desire on the part of the former to maintain the metering apparatus in first-class condition. As may be inferred from the preceding discussion, if meters are to be depended upon for accuracy and reliability, much greater vigilance will be required than is the case with electric meters. This requires a thorough inspection service, combined with close scrutiny of the records of steam consumption to determine if the meter is operating satisfactorily. The causes of faulty registration may be most conveniently grouped under the following heads, viz.:

GROUP I.—METER DEFECTS

Area meters.—Meters may be too large for the average load. Valve may stick or bind due to foreign matter in steam. Orifice may become enlarged if valve pounds on seat. Clocks and recording mechanism may be out of order on account of heat, vibration or other causes. Pencil arm may be loose on shaft. Meter out of vertical alignment.

Velocity meters.—Pitot-tube may be twisted out of position in the pipe. Pitot-tube may become partially clogged by foreign particles in steam. Leaks in piping from tubes to manometer. Loss of mercury due to leaks. Loose wiring connections to recording instruments. Faulty voltage regulation for recording instruments.

Condensation meters.—Meter too small or improperly installed. Tight bearings —need of lubrication. Oil or other foreign matter in drums or buckets. Spinning or sticking of drums or buckets. Meters out of level. Tampering by consumer. Nozzle or spout obstructed. Moving parts out of their proper bearings. Register trouble—dial hands and gears loose, pinions bent, etc.

GROUP II.—SYSTEM TROUBLES

Installation.—Vapor vents to relieve excess pressure. Leaks in steam piping. Leaks in return piping. No protection against surging or flooding. Wrong type of steam trap. For vacuum meters a vacuum pot or trap is necessary. Not sufficient fall to the sewer. Not sufficient head on the inlet to secure full capacity of meter. Sediment trap should be provided.

Operation.—Defective and leaky steam traps. Sewer clogged up. Condensation diverted, intentionally or otherwise. Flooding of meter in early morning, due to returns being held up during the night. Sediment trap filled. Chemicals in return water may corrode parts of meter. Temperature too high.

A study of the above emphasizes the great importance of this phase of the subject but unfortunately lack of space will not permit of a complete discussion of the points referred to. The literature to be procured from the manufacturers will call attention to the care necessary for successful operation of each individual meter, but the reader is urged to consult the " Transactions of the National District Heating Association," 1912–13–14, in which volumes a large fund of valuable data is to be found. These data have been prepared by committees appointed by the Association, and their reports, together with the discussions of same form the most authentic account of the experiences of many steam heating companies. A digest of these reports brings out the following important points:

Meters should be read as often as practicable in order that any defect may be immediately noted and rectified. It is not sufficient to follow the policy of the electric companies, where meters are read monthly. Much more attention is necessary.

Meters should be tested at least once each year and oftener, if erratic records have been obtained.

Much of the trouble is wrongly charged to meters since it may be the result of defects in the system of piping.

Where meters are not used in the summer season they should be protected against corrosion by being properly drained and painted.

Meters are more liable to under-register than otherwise, therefore the expense of maintaining them in good condition is in the interest of the company no less than the consumer.

GENERAL CONCLUSIONS

It might be thought within the scope of this work to differentiate more closely as to the relative accuracy, and reliability of the different meters which have been described, but in view of the fact that the actual development of the leading makes of meters, has covered only a very few years and practically all of them are now undergoing refinement in various degrees, it is obviously unfair to pronounce final judgment upon the question.

In general those manufacturers who recognize the paramount necessity of calibration are most likely to produce a device which can be relied upon to fill the requirements. Calibration is the key-note of the art; and although we have covered at some length the theoretical principles involved and found that the various formulas could be applied with fair approximation to practical results, yet commercial requirements demand some assurance that the instrument will have been thoroughly tested under somewhat similar conditions to those for which it is to be used. A considerable part of the cost of steam flow-meters is due to the expense of calibration but from this there is no escape. With this done, the steam meter may be considered as an instrument of convenience and practical utility, although in the present state of development it perhaps lacks many of the elements necessary for universal adoption.

There are few, if any, decisions which prescribe the maximum allowable error for steam companies in their dealings with the public. After considering all the facts, it would seem that commercial steam-service should be supplied on the basis of an allowable variation of from 5% to 7% from the actual consumption. In other words, the companies should not be under strict legal requirements to adjust bills, unless the error has been established as lying beyond this limit, although the calibration constant of the meter should be immediately corrected according to the test results. This practice would agree with that adopted by many electric companies, although the limit of error is somewhat lower, generally 4%, which is to be expected on account of the greater ease of measuring electricity.

With reference to condensation meters, it has been seen that it requires only attention to the details of installation to procure uniform results and place these devices on a par with electric meters, as regards simplicity and precision. When choosing between the different meters now on the market it is well to rely upon the evidence to be procured from those steam companies which have experimented upon all the various types. The selection of a meter should be governed by its general reliability and expense of maintenance; as well as upon its accuracy and initial cost.

The drawback connected with condensation meters is the possibility of not receiving all the returns from the system. These may be diverted in many ways by unscrupulous customers, or through leaky piping which often escapes inspection. Frequently, it is necessary to supply steam for open jets in restaurants and industrial plants and it is impossible of course to meter the steam used for these purposes by the condensation method. Nor can these meters be used where any steam pumps are operated, unless some method is adopted of condensing all the exhaust steam. Thus it is seen that the utility of this device is confined within certain limits which prevent its universal adoption as a substitute for the steam meter, although it will probably never be supplanted for small consumers of steam.

The ideal meter, perhaps, is one which will measure the rate of flow of steam as it enters the consumer's premises and will automatically integrate the quantity in a way similar to that which prevails in the measurement of other domestic utilities—gas, water, and electricity. Such a device should be very simple in construction so as to keep the cost within reasonable limits and confine the number of parts to a minimum. The recording feature should be retained, as it furnishes indisputable evidence in case bills are questioned. It is of course, desirable that the fluctuations in pressure be compensated by some simple method. If these points were incorporated into an instrument the scientific conduct of district heating would be an assured fact.

Attempts have been made to build satisfactory instruments for the purpose of metering the heat supplied in terms of British thermal units, instead of the prevailing custom—pounds of steam. Such a device is urgently needed by hot-water heating companies as by this method the present compulsory use of flat rates would then be unnecessary. No substantial progress has been made along these lines, which fact attests the difficulties in the inventor's path. The velocities employed are relatively low in hot-water distribution systems thus eliminating consideration of the Pitot-tube. Some modifications of the Venturi-tube might be suggested in conjunction with a device for recording the drop in temperature; or some improved form of displacement or volumetric meter might be adopted with greater chances of success. Considering the relatively low differentials of pressure and temperature, sensitive instruments would be imperative and the consequent cost might prevent the success of such a meter.

A general survey of the subject of steam measurement forces the conviction that while much has been accomplished there are still vast areas comparatively unexplored. American inventors are constantly working on the problem but they are not alone in their endeavors. To the Germans a large share of credit is due for their investigations and labors in the same field. The well-known proclivity of the German scientific mind for patient and persistent study has been and will be of enormous benefit in solving many of the exacting problems connected with the metering of the steam.

CHAPTER V

DISTRICT HEATING STATIONS

In all district heating-stations, whether the service is supplied by means of hot water, or by a steam system, the primary source of heat supply is the furnace and boiler. It will therefore be interesting to take a brief survey of the different types of boilers and furnaces now in use, with a few notes as to the advantages of the different types used.

TYPES OF FURNACES

Furnaces for the combustion of fuel are made in various types. They may be classified in tabulated form as follows:

(a) Hand fired
- Ordinary stationary or straight grates;
- Rocking, shaking and dumping grates;
- Same as above with fire brick, or tile arches, or baffles, for more complete combustion and smoke prevention, built under the sheets or tubes of the boiler
- Extended furnaces of the Dutch-oven type built out in front of the boiler and usually fed from the side, such as the Burke furnace.
- Down draft furnaces

(b) Stoker fired
- Chain Grates
  For example—Green, Babcock & Wilcox, Illinois, Playford, LaClede Christy and MacKenzie
- Inclined Grates
  - Side Inclined — Murphy, Detroit, Model
  - Front Inclined — Roney, Ross, Wilkinson
- Underfeed
  - Horizontal — Jones, American
  - Inclined — Taylor, Westinghouse

(c) Oil burners
- Compressed air atomizers.
- Steam atomizers
- Mechanical or pressure atomizers

(d) Gas burners

(*a*) **Hand Fired Furnaces.** The advantages of hand fired furnaces are:

1st. Ease of manipulation. With hand fired furnaces, the fire can be supplied with coal at almost any rate that is required and the furnace can be watched very closely. In this way the amount of coal fed into the boiler can be made to correspond very closely to the steam requirements. The disadvantage of the hand fired furnace is that it requires more careful attention on the part of the fireman and offers many opportunities for waste of coal by improper firing. Another obvious disadvantage is the increased cost for labor in large plants.

(*b*) **Stoker Fired Furnaces.** Under stoker fired furnaces, we have three main divisions. All three types of stokers have their champions and advocates. The choice of a stoker depends upon the kind of coal to be burned. The chain grate-stoker is best adapted for free burning bituminous coal in the smaller sizes, such as $1\frac{1}{2}$ in. screenings and No. 4 washed coal. This type of stoker has many advocates but its use is generally adapted to plants having a uniform load condition. It is therefore especially to be recommended for heating stations since in this class of service the variations in the load can be more easily anticipated than is practicable in some electric lighting plants.

Stokers of the inclined grate type are best adapted to coking coals, although they may be used for the smaller sizes of bituminous coals. They respond more readily to fluctuating load conditions than chain grate-stokers, but require more manual labor on account of the necessity of removing clinkers. These stokers will operate with either No. 4 or No. 5 washed coal or $1\frac{1}{2}$ in. screenings. A selection between these three grades of coal will depend upon the rate at which it is necessary to burn the coal, and also on the particular make of inclined grates used.

The underfeed stokers are employed with greatest success with coking coals. Where a very good grade of coal can be obtained, which is suitable for burning by the retort method and forced draft, satisfactory results can be expected. The best coals for this type of stoker are Pennsylvania and Virginia semi-bituminous. These stokers are sometimes used, however, with inferior grades of coal, although the liability of clinkering of the ash constituents in the coal interferes with their best efficiencies. The size of coal usually used with this type of stoker is No. 4 washed coal or $1\frac{1}{2}$ in. screenings.

Mechanical stokers burn fuel in a more economical manner than is found possible under the average hand fired method. This is due to the fact that the coal may be fed continuously in small quantities of uniform thickness whereas with hand firing a large quantity of coal is introduced into the furnace at irregular intervals, resulting in poor combustion and smoke production. To partially offset the saving in labor and increased economy, stokers possess the disadvantage of requiring considerable space, this being true especially with chain grates. Furthermore, a complete stoker installation for a modern plant must be equipped with coal and ash handling systems, which add to the initial investment and also to the cost of repairs, and for plants of less than 500 to 800 horse-power, it is usually unadvisable to install them. In the large plants and particularly in localities where coal and labor are expensive the economy of mechanical stokers becomes very evident.

(c) **and** (d). **Oil and Gas Burners.** The use of oil and gas is confined to certain localities where the supply is cheap and abundant, but where possible, oil fuels offer many advantages over coal, among which may be enumerated the following:

1. Less space occupied for equivalent heat capacity.
2. Cost of labor and handling much lower.
3. Higher combustion efficiency, and smokelessness.
4. Cleanliness of heating surface.

Certain disadvantages are found with the use of oil fuels, but they are mainly mechanical and may be eliminated by careful engineering. The great majority of steam generating-plants, with the possible exception of those in the Pacific Coast territory, are now using coal on account of the fact that the supply of oil or gas is likely to be uncertain.

Ordinary definitions of the different sizes of coal. Different coal dealers figure somewhat differently as to the size of bituminous washed coal. For the purpose of this chapter, the commercial sizes of coal will be assumed to be those sizes which will pass over certain sizes of screens and which will pass through certain screens, as follows:

Size of coal	Will pass over	Will pass through
No. 1	$1\frac{3}{4}$ in. screen	$2\frac{1}{2}$ in. screen
No. 2	$1\frac{1}{8}$ in. screen	$1\frac{3}{4}$ in. screen
No. 3	$\frac{3}{4}$ in. screen	$1\frac{1}{8}$ in. screen
No. 4	$\frac{1}{4}$ in. screen	$\frac{3}{4}$ in. screen.

No. 5 size contains all coal that will pass through a $\frac{1}{4}$ in. screen, with the exception of very fine duff which is supposed to be removed in the washing. $1\frac{1}{2}$ in. screenings are generally considered as coal which will run from very fine coal up to $1\frac{1}{2}$ in. in size with scattered pieces of coal which are somewhat larger.

BOILERS

The above types of furnaces can be used with almost any type of boiler. The following in tabulated form are the different types of boilers most commonly used:

(a) Water-tube boilers
- Horizontal pass boilers   Heine, Edgemoor, Geary, Franklin, O'Brien, Keeler and Vogt.
- Vertical pass boilers.   Babcock & Wilcox, Stirling, Rust, Morrison, Springfield, Wickes, Winslow, Union.
- Combination of horizontal and vertical passes in boilers.   Almy, Badenhausen, Worthington, Freeman.

(b) Fire-tube boilers
- Ordinary return tubular boiler, externally fired.
- Scotch Marine type internally fired { Continental Freeman.

(c) Combination water tube and fire-tube boilers { Bonson, Sederholm, Hawkes and Lyons

Many of the above water-tube boilers may be changed from horizontal to vertical pass and vice versa merely by rearranging the baffles. As far as efficiency is concerned, there is probably little choice between the fire-tube and water-tube boiler, except that water-tube boilers may be designed for a higher steam pressure, which results in a greater efficiency in large plants.

The cost of the fire-tube boiler is usually less than that of the water-tube boiler, but on the other hand, the space required for the water-tube boiler is usually less than that required for the fire-tube boiler. The fact that in the water-tube boiler, the steam is confined within the tubes instead of being in large shells reduces the danger of boiler explosions, and therefore from the standpoint of safety, the general practice has turned largely to the use of water-tube boilers.

Auxiliary Power Plant Appliances. One device frequently advocated for power plants is that known as the " Economizer," which consists of pipes or coils placed in the breeching, leading to the stack and which gives additional heat to the boiler feed-water supply before it reaches the boiler. The objection to economizers is that they take up considerable space and reduce the available draft over the fire. They sometimes require the use of forced draft in order to overcome the obstruction. The tubes need considerable attention in order to remove the deposits of soot and ash as well as for the removal of scale from the interior of the tubes.

It is, therefore, necessary beforehand to figure very carefully as to whether the cost of installation and maintenance of fan blast will not offset the added efficiency and economy in producing steam. In the small boiler plants, this will be found to be the fact, but in very large plants, the economizer has been tried out with considerable success.

Another apparatus, for use only in by-product or exhaust steam heating-plants is the super-heater. By means of this, the steam after leaving the boiler is brought into a receptacle placed in the path of the furnace gases where additional heat is imparted to the steam, thereby increasing the temperature above the normal. Superheated steam is rarely used by heating companies.

There are also instruments for the purpose of showing the relation between the gases of combustion and the amount of air fed into the furnace. These may be used as a guide to the engineer in regulating the dampers for the supply of air.

Some Points in Regard to Economical Operation. The economical operation of heating plants is a complex study. There are various factors which enter into the problem, among which the following might be enumerated:

1. The necessity of a constant supply of coal suitable for the use of the particular boiler and furnace equipment employed.

2. The employment of a size of boiler or boilers approximately suited to the requirements of the load.

3. The arrangement of the plant, so that at all times there may be a reserve boiler which is in process of cleaning, while the other boilers are in operation.

4. The importance of the removal of the accumulations of soot and ash from the outside of the boiler tubes and also of scale from the inside of the tubes cannot be overestimated. No matter how thorough the combustion of the coal and no matter how well proportioned are the furnace gases, it will be impossible to secure economical operation, unless the heat from the furnace gases finds free passage through the boiler tubes to the water and steam.   Iron and steel are very good conductors of heat, but soot and scale are very poor conductors.

To illustrate, a certain plant formerly required four 500-horse-power boilers to keep up the load during the winter season, and as there were only four boilers installed this meant little opportunity for cleaning the boilers, during the cold winter months. The plant has since been thoroughly overhauled, with the result that the same load may be carried with three boilers that was formerly carried with four, and with greater ease. By frequently changing the boilers, and replacing one of the used boilers by a boiler thoroughly cleaned, the plant has been kept in efficient condition throughout the winter. This point of cleaning applies not only to the boiler tubes themselves, but also to the hot-water heater for feed water and other hot water requirements.

The central heating-plant is usually better able to use boilers which are of suitable size for the required load.   The individual plant frequently has a load in the summer time which is hardly appreciable, but at the same time a large boiler is kept under steam in order to heat a few barrels of water and a few steam tables in some restaurant.   The result is extreme inefficiency.

In a certain plant a 500-H.P. boiler was employed during the heating season to supply less than 400,000 lbs. of steam per month, with a result in efficiency of less than one pound of steam per pound of coal burned. Had this plant been connected to a number of other plants in accordance with the group system of heating, there would have been a very material saving both in coal and labor. On the other hand, in the winter time it is frequently possible in the group system of plants to take the large boiler plant and heat a group of buildings from a single plant, thus doing away with the operating of several small plants and their respective quota of men.

Another advantage of the district heating-system is its ability to employ a higher grade of engineers than the ordinary small plant can afford. The ordinary building owner hires an engineer at a salary of from $75.00 to $80.00 per month and expects him to know all about the theory of the boiler plant, the theory of the machinery in the building and understand all points in connection with the circulation of the steam through radiators and piping. In other words, he expects a grade of intelligence for which he would pay from $150.00 to $200.00 per month in other departments of the business. It is needless to say that he is usually disappointed and the fact that he does not realize the waste that is going on in his establishment on account of his own ignorance is no sign that the waste is not going on.

A careful inspection of some of the individual plants will cause the intelligent observer to wonder that there are not more boiler explosions and accidents than actually occur, when he sees the " hit or miss " methods and the large use of guess work employed around the ordinary small plant. Frequently a little careful and intelligent supervision will decrease the operating cost of heating plants from 10 to 25%. When to this is added the increased efficiency through the operation of large units, there is found a substantial saving through the operation of a district heating company.

As stated above, one of the primary requisites in the operation of a plant is a constant supply of coal adapted for use in the particular boiler employed. This can be secured only by an intelligent purchase of coal. The average coal dealer is apt to advocate the type of coal he has on hand, rather than the coal which is suitable for the requirements of the customer. He is seldom interested in the efficient handling of furnaces, inasmuch as the more inefficiently the furnace works the greater will be his sales of coal.

To a great many building owners, coal is coal, just the same as steam is steam and they fail to differentiate between the different values of various grades of coal. Even if a building owner understands that some coal is more valuable than others, he is usually at a loss to know just what to ask for or what to expect. By a comparatively small investment, a laboratory can be equipped in which may be analyzed the various classes of coal delivered and accurate results obtained as to the heat value of the coal. The following is a list of apparatus and its cost as used in a private laboratory in Chicago:

1 Parr Calorimeter, (hot wire)	$70.00
1 thermometer, graduated from 65° to 105° F., $\frac{1}{20}$°.	15.00
1 balance, reading to $\frac{1}{10}$ of a millimeter	58.50
1 box weights, from 1 mm. to 50 gms.	9.00
1 Bunsen burner	.55
1 desiccator, 6 in.	2.45
1 crucible porc. size oo, with cover	.22
1 pair crucible tongs	.35
1 watch glass, $2\frac{1}{2}$ in. diameter	.05
1 sieve, 100 mesh, 5 in. diameter	1.40
1 triangular crucible holder	.09
1 copper oven, 6 in. ×8 in.	4.00
1 thermometer	1.00
1 grinding board and rubber	7.00
1 brush, $4\frac{3}{4}$ in. wide	.70
1 flask, 200 cc.	1.00
1 jar, sampling glass	.25
1 bottle, 2 oz.	.10
1 spatula, 5 in.	.25
1 lb. sodium peroxide for Parr's calorimeter	2.00
1 bottle accelerator	.50
2 lbs. calcium chloride, for desiccator	.70
Slow speed, $\frac{1}{8}$ H.P. motor	25.00
	$200.11

The above list totals a trifle over $200.00 for the entire investment. The list of apparatus is based on the use of a Parr calorimeter which is not quite as accurate as the Mahler bomb calorimeter, but is sufficiently accurate for practical checking purposes. A simple method of analyzing coal is as follows:

Sampling. In sampling coal for the purpose of analysis every precaution should be taken to have the sample fairly representative of the entire quantity of coal as there is more chance for error in improper sampling than in the analysis. In order to obtain a fair sample of coal, at least 150 to 200 lbs. should be taken from the top, center and bottom of the pile in such a manner as to get a fair proportion of the large and small sizes. The sample should be placed on a clean, hard dry surface and all the large pieces broken up so that all will be about the same size. After this is done, the sample should be thoroughly mixed with a clean shovel, and made into a round pile or pyramid.

Then the pile should be carefully divided into equal quarters, and the two opposite quarters thrown away. After this has been done, the fine coal remaining on the floor should be carefully swept away with a clean broom, leaving the floor in the same condition as it was before the sample was placed upon it. The two remaining quarters should then be thoroughly mixed, pounded so as to further reduce the larger pieces and quartered as before. This quartering, pounding and mixing process should be continued until there is about a pint left, none of the pieces of coal being larger in size than a grain of wheat after the final quartering. The sample should then be put into an air-tight can and a label placed on the sample can giving the dealer's name, kind of coal, date, by whom collected, and where sample was taken. The whole process of sampling should be performed as rapidly as possible, so as not to lose any of the moisture contained in the sample.

Determination of Moisture. The sample is then delivered to the laboratory, recorded in the laboratory data book, and given a number. The contents of the can are emptied on a square piece of oil cloth and five grams is taken and weighed very accurately on a sensitive balance which will weigh $\frac{1}{10}$th of a milligram. The five-gram sample of coal is then placed on a watch-glass which has the same laboratory number as the sample and is then put into an oven at a temperature from 220° to 230° F. to evaporate off the moisture. The five-gram moisture sample is dried under these conditions for an hour and a half. It is then removed from the oven, cooled in a desiccator over calcium chloride, and when cooled is again weighed. The difference between the original and the final weight multiplied by 100 and divided by the original weight gives the per cent of moisture contained in the sample.

Determination of Ash. The balance of the sample is then placed on a grinding board and is ground down until it will pass through a forty-mesh sieve. It is again mixed and quartered until there are about two ounces left, after which it is again ground until it will pass through a 100-mesh sieve. It is again mixed and placed in a two ounce wide-mouthed bottle with a ground stopper. This bottle is numbered and dated. About 5 grams of this finely-ground coal is then taken from the bottle and placed on a watch-glass in a manner similar to that pursued when making the moisture determination. It is then placed in an oven, and kept at a temperature of 220° to 230° F. for an hour. After being thus dried, it is then cooled in a desiccator over calcium chloride. When cooled one gram is weighed out accurately and put in a porcelain crucible with a porcelain cover, which is then placed over a Bunsen burner and slowly heated until the volatile matter in the coal is driven off. The object of this slow heating is to avoid coking the coal. When the volatile matter is driven off the burner is turned up so as to get a high temperature. After the coal has remained on the burner for one hour it is stirred with a platinum wire until all particles of carbon have disappeared. It is then left to burn for one-half hour more so that all combustible matter will be thoroughly ignited. After being taken off the burner it is placed in the desiccator to cool, after which it is again weighed. This weight multiplied by 100 and divided by the original weight will give the per cent of dry ash.

Determination of British Thermal Units. In obtaining the number of B.T.U. per pound of coal, it is necessary to have perfect combustion and complete absorption of the heat produced. This is done by means of the Parr calorimeter. The coal to be tested for heat value must be perfectly dry and ground down so that it will pass through a 100-mesh sieve. Exactly one-half gram of this coal is placed in the cylinder of the calorimeter, with about nine grams of sodium peroxide to supply oxygen needed for combustion; and to secure more rapid and complete combustion, one gram of a mixture of two parts potassium persulphate and one part ammonium sulphate is added. After the coal and chemicals have been carefully put into the cylinder, it is closed and securely fastened.

Then it is well shaken to mix the coal and chemicals thoroughly and the cylinder is lightly tapped to shake all the material from the upper part of the cylinder. A pair of propeller blades are attached to the cylinder for the purpose of keeping the water in circulation and maintaining a uniform temperature throughout the water. The cylinder is now placed in a can, which contains 2000 cubic centimeters of clear water. The water is brought to a temperature of about 2° F. below that of the laboratory, so that the average temperature of the water during the rise of temperature due to igniting the coal will be about the same as the room temperature, and inaccuracy will be reduced to a minimum.

A thermometer is now placed in the water so that the mercury bulb will be about midway between the top and bottom of the can of water. The cylinder is next connected to a small motor by means of a pulley and belt and made to revolve at about 100 R.P.M. After about three or four minutes the first reading of the thermometer may be taken. The mixture is then ignited with a red hot piece of wire, dropped through the stem of the cylinder resulting in a rise in temperature within the cylinder which still remains in motion. The heat is transmitted to the water in the calorimeter and thermometer readings are taken every few seconds until the highest reading is obtained. To obtain the B.T.U. per pound of dry coal, subtract the initial temperature readings from the highest temperature recorded. From this observed temperature difference a correction, due to the heat liberated by the ignition of the chemicals and heating of the iron wire, must be subtracted and a further allowance must be made for the ash and sulphur present in the coal. The correction factors for a Parr calorimeter are as follows:

CORRECTION COMPUTATION

For ignition by hot wire.............. 0.022° F.
For each per cent of ash............. 0.005° F.
For each per cent of sulphur......... 0.010° F.
For 1 gram of mixture of potassium per-
  sulphate and ammonium sulphate... 0.234° F.
For coal having volatile matter over
  30% include a further correction.... 0.045° F.

For example, in analyzing a coal containing 9.0% ash and 2.0% sulphur, the summary of the above correction would be as follows:

Hot wire.......................... 0.022°
Dry ash........................... 0.045°
Sulphur........................... 0.020°
Chemicals......................... 0.234°
Volatile matter................... 0.045°

    which equals.................... 0.366°

Observed Thermometer rise........... 4.36°
Subtracting correction.............. .366°

Net Thermometer rise................ 3.994°
This multiplied by the Calorimeter con-
  constant, e.g...................... 3,117
Gives the dry B.T.U. per pound of dry
  coal, or.......................... 12,449 B.T.U.

COAL COMPUTATION

Determine the per cent of dry coal by subtracting the per cent of moisture from 100%.

Determine the commercial B.T.U. by multiplying the dry B.T.U. by the per cent of dry coal expressed as a decimal.

Divide this commercial B.T.U. by one-half the contract guarantee or one-half of the B.T.U. guaranteed for one cent and multiply the result by ten to express it in dollars.

This is the same as multiplying the commercial B.T.U. by 2000 to get the B.T.U. per ton, then dividing by the contract guarantee.

From this value subtract one-half the delivered dry ash percentage expressed in cents, this giving the delivered value of the coal.

To determine the delivered B.T.U. for one cent, multiply the commercial B.T.U. by 2000 to find the B.T.U. per ton of coal. Add one-half the dry ash percentage expressed in cents to the contract price and divide the B.T.U. per ton by this figure.

During the past few years a great deal of interest has developed in the purchase of coal on the basis of specifications which specify the quality of the coal to be furnished and heat value of same. The United States Government has adopted a set of specifications and buys coal almost exclusively on the heat valuation of the coal. Extensive tests have been made on coals from different localities and information on these classes of coals is much more general than it was a few years ago. A number of bulletins and pamphlets have been issued by the Department of the Interior, Bureau of Mines of the United States Government, which give valuable and instructive information in relation to the analyzing of coal and burning of coal in different types of furnaces. A great many of these bulletins are furnished free on application to the Government, while for others a slight charge is made. Among the interesting bulletins may be mentioned:

Technical papers, No. 8 and No. 76.
Bulletins, Nos. 41 and 63.

The following is a typical form of contract which has some-times been used in the middle west for the purchase of coal:

" THIS AGREEMENT, made and entered into by and between
. . . .
. . . .
. . . . .hereinafter called the " Consumer," and. . . . .
. . . .
hereinafter called the " Company," Witnesseth that;

" *Whereas*, it is desired by the parties hereto to enter into a contract for the purchase and sale of coal upon the heat value basis, as herein set forth:

" Now, therefore, in consideration of the mutual covenants and agreements herein contained, the parties hereto do hereby agree as follows:

" 1. The Company agrees to furnish and deliver to the Consumer. . . . .
. . . .
at such times and in such quantities as ordered by the Consumer for consumption at said premises during the term hereof, at the Consumer's option, either or all of the kinds of coal described below:

Kind of Coal:. . . . .

" Of size containing all of the coal that will pass through a screen having circular perforations of the following diameter in inches. . . . .

" Or through a diamond bar-screen having the following clear distance in inches between the bars. . . . .

" And containing not more than the following per cent of coal by weight that will pass through a screen having circular perforations one-quarter of an inch in diameter. . . . .

" Said coal to be mined from Geological seam number. . . . . . .

" In the following county. . . . .

" It being agreed that said coals are to average the following assays:

" Per cent of moisture in coal as delivered. . . . .

" Per cent of ash in dry coal. . . . .

British Thermal Units per pounds of dry coal. . . . .

" Coal of the above respective descriptions and specified assays (not average assays) to be hereinafter known as the con-tract grade of the respective kinds. If more than one kind of coal is described above, the Consumer shall notify the Company in writing within thirty days after date hereof, which kind or kinds of coal and what proportion of each, it wishes to receive. If no notice is sent as above, it is understood that the Company shall deliver only the kind of coal described by the left-hand column of data.

" II. The Consumer agrees to purchase from the Company all of the coal required for consumption at said premises during the term of this contract except as set forth in Paragraph III, below, and to pay the Company for each ton of 2000 pounds avoirdupois of coal delivered and accepted in accordance with all of the terms of this contract at the following contract rate per ton for coal of each respective contract grade, at which rates the Company will deliver the following respective numbers of Net British Thermal Units for one cent, the contract guarantee:

Kind of Coal.	Contract Rate per Ton.	Contract Guarantee.
. . . .	$. . . . . . Equal to.	Net B.T.U. for one cent.
. . . .	$. . . . . . Equal to.	Net B.T.U. for one cent.
. . . .	$. . . . . . Equal to.	Net B.T.U. for one cent.

" Said Net British Thermal Units for one cent being in each case determined as follows: Multiply the number of British Thermal Units per pound of dry coal by the per cent of moisture (expressed in decimals); subtract the product so found from the number of British Thermal Units per pound of dry coal; multiply the remainder by 2000 and divide this product by the Contract Rate Per Ton (expressed in cents), plus one-half of the ash percentage (expressed in cents).

" III. It is provided that the Consumer may purchase for consumption at said premises coal other than herein contracted for, for test purposes, it being understood that the total of such coal so purchased shall not exceed five per cent of the total consumption during the term of this contract.

" IV. It is understood that the Company may deliver hereunder coal in which the ash percentage is as high as three per cent more, and in which the moisture percentage is as high as five per cent more, and in which the percentage by weight of coal which will pass through a screen having a circular mesh one-quarter of an inch in diameter is as high as five per cent more, and in which the number of British Thermal Units per pound, dry, is as low as 500 fewer than specified above for contract grade.

" V. Should any coal delivered hereunder contain more than the per cent of ash or moisture, or fewer than the number of British Thermal Units per pound, dry, allowed under Paragraph IV hereof, the Consumer may, at its option, either reject or accept same. Should any coal delivered hereunder contain more per cent by weight of coal which will pass through a screen having a circular mesh one-quarter of an inch in diameter than is allowed under Paragraph IV herein, the Consumer may, at its option either reject same or accept same and deduct from its delivered value as a penalty, an amount equal to one and one-half per cent of the Contract Rate Per Ton for each one per cent so in excess of the percentage allowed in Paragraph IV hereof.

" VI. All coal accepted hereunder shall be paid for during the calendar month following delivery at a price per ton determined by taking the average of the Delivered Values obtained from the analyses of all the samples taken during the preceding month, said Delivered Value in each case being obtained as follows: Multiply the number of British Thermal Units delivered per pound of dry coal by the per cent of moisture delivered (expressed in decimals); subtract the product so found from the number of British Thermal Units delivered per pound of dry coal; multiply the remainder by 2000 and divide this product by the Contract Guarantee; from this quotient (expressed as dollars and cents) subtract one-half of the dry ash percentage delivered (expressed as cents).

" VII. For the purpose of determining the quality of coal delivered hereunder, it is agreed that the Consumer shall cause samples to be collected and analyzed by the . . . . at the Consumer's expense, as follows: A fair average sample of the coal delivered hereunder shall be collected not less than once each week. If only one such sample is collected each week, each such sample shall be analyzed. If more than one sample is collected per week, composite samples shall be made semi-monthly and analyzed. Each such composite sample shall include all of the samples collected since the previous composite sample was made. The results of all analyses made, together with the Delivered Values determined by such analyses, shall be reported promptly to the Company. The Company may have a representative present at the time of selecting of any or all samples as above, and when requested by the Company in writing, such sample shall be divided into three parts, one given to the Consumer, one to the Company and the third sealed in the presence of the representatives of both and kept by the Consumer. If the Company requests, in writing, within five days after receipt of the report of the Consumer's analysis, as provided above, the third sealed part of the sample shall be delivered to a chemist of high standing in fuel analysis, to be mutually agreed upon, and analyzed at the expense of the Company, and this analysis shall then be final and binding.

"VIII. The Company agrees that, in the event of a strike or strikes preventing all deliveries of coal of contract grade, it will notify the Consumer in writing of said strike or strikes and will deliver to the Consumer, coal of quality and price equivalent to that of the contract grade in sufficient quantity to meet the Consumer's requirements for thirty days after said strike or strikes take place.

"Should any such strike or strikes continue and prevent all deliveries of coal of contract grade for a longer period than thirty days, it is agreed that the Consumer shall then buy, exclusively from the Company, such coal as it requires, of quality equivalent to that of the contract grade providing the Company will sell such coal at as low a price as the Consumer can obtain it for elsewhere. If the Company does not offer to sell such coal at as low a price as the Consumer can obtain it for elsewhere, then the Consumer may purchase such coal elsewhere. In this event the Consumer's cost of coal, in excess of the cost of an equivalent amount of coal of contract grade hereunder, shall be borne equally by the Company and the Consumer.

"When said strike or strikes are over, or delivery of coal of contract grade can again be made, the rights and duties of the parties hereto shall be the same as before such strike or strikes occurred.

"IX. Should the Company fail to deliver coal in accordance with the terms of this contract, the Consumer may, at its option, cancel the then unexpired portion of this contract, or any portion thereof, or the Consumer may purchase on the open market such an amount or amounts of coal as the said Company shall fail to deliver, in which events all costs and expenses occasioned by such failures, in excess of the cost hereunder of an equivalent amount of coal of corresponding grade, shall be borne and paid by said Company.

"X. When requested by the Consumer, the Company shall show the Consumer the original bills of lading of cars from which coal is delivered hereunder.

" XI. The term of this contract shall be from the date hereof to and including the..............day of..................

In Witness Whereof, the parties hereto have hereunto set their hands and seals, at................, this......day of....
..................19...

....

....

ATTEST:

....

....

In the operation of the above contract, the value of the coal delivered each month would be determined by a number of analyses and the price would go up and down, according to the average results shown by these analyses. This type of contract has sometimes been the cause of considerable controversy between coal companies and customers, inasmuch as a slight difference in one or two analyses often affects the price of a large amount of coal.

In order to protect the consumer against poor coal and at the same time not to penalize the coal companies for various amounts of moisture found in coal during winter weather and slight changes over which they have no control, the following modified clauses were adopted by a large western heating company and have been used with considerable success. As long as the coal is approximately the quality contracted for the coal company is not penalized, but when it falls below a certain standard in any month, the price of the coal is based on the analyses of the coal. The following are the special clauses used in this contract:

" All coal furnished hereunder shall be paid for at the regular price per ton, except when the delivered value drops ten cents per ton below the above guaranty. In such cases multiply the number of British Thermal Units per pound of dry coal delivered by the per cent of moisture (expressed in decimals), subtract the product so found from the number of British Thermal Units per pound of dry coal: multiply the remainder by 2000 and divide this product by the net British Thermal Units for one cent of the respective coals. From this quotient (expressed as dollars and cents) subtract one-half of the ash percentage (expressed as cents).

" It is understood that the analysis of coal is not to be used in order to regulate the price from month to month, but is only to be used when inferior coal has been delivered, and in such cases the amount of reduction to be made on account of the inferior quality of the coal shall be determined by such analysis."

Another point to be considered in the purchase of coal is the advantage of providing for a certain amount of storage coal during the winter months. The amount of storage coal varies in different localities. On an average this should be about 5% of the total year's requirements. This amount of storage is usually only required when the demand is at its maximum. Theoretically the railroads should be able to deliver in cars, day by day, coal in accordance with requirements; but as a matter of practical experience the coal deliveries are frequently delayed for days and sometimes weeks, and in order to have a supply of the different kinds of coal for each plant, it is necessary that storage coal be provided to be drawn upon in emergencies.

Smoke Prevention Devices. It is in accordance with the tendency of the times that heating plants be required to operate without smoke. In this connection there is found a great advantage in district heating-stations. In the case of the small hand-fired boiler plant, the owner is often at the mercy of an ignorant fireman as far as his ability to keep down the smoke is concerned. In many cities the emission of dense smoke from the chimneys of office buildings and manufacturing concerns, has been made a quasi-criminal offense. Thousands of dollars in fines have been collected from plant owners due to their inability to determine the cause of smoke and to accomplish its prevention.

Smoke is nothing more than fine particles of carbon which have not been consumed in the boiler plant and is, therefore, an index of improper combustion. It is not only a nuisance to the general community but also is evidence of inefficient plant

operation and waste of coal. Various schemes have been
suggested for the prevention of smoke. A great many of these
are based on different types of baffling in the boilers, these
being instrumental in effecting complete combustion by preventing contact of the flames with the relatively cool boiler
tubes. A number of different types of boiler settings are
shown in the accompanying illustrations.

Fig. 54 shows the arrangement of furnace and boiler in a
large office building. The furnace is of a chain-grate type and

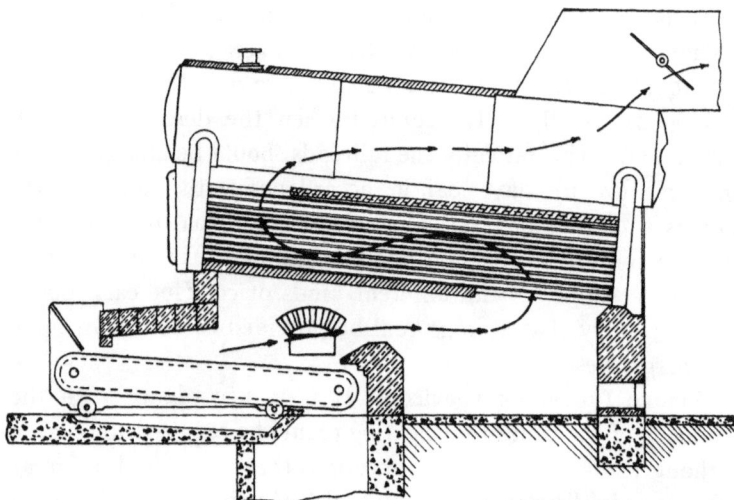

FIG. 54.—Heine Horizontal-pass boiler—Illinois stoker

is used in connection with Heine boilers. This arrangement
of furnace and baffling has given very good results, the plant
having been operated several years with practically no smoke.

Fig. 55 shows the Scotch-Marine boiler with Burke furnace.
This type of furnace is a Dutch-oven arrangement and has
been used for a number of years with satisfactory results.

Fig. 56 shows a combination of the Babcock & Wilcox
boiler with the Roney stoker, used in the plant of a large office
building. This arrangement of baffling while suitable for
efficient production of steam is not a good one from the stand-

point of smoke consumption. In the particular plant in which this arrangement is found, it is impracticable to use the hori-

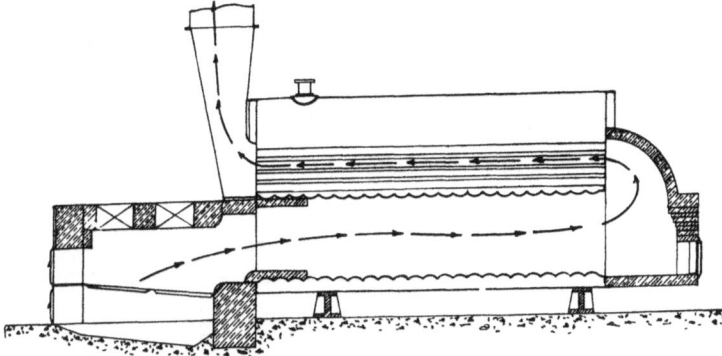

FIG. 55.—Internally-fired boiler—Burke furnace

zontal baffling, due to the difficulty in getting at the tubes for cleaning purposes, and also to lack of sufficient draft. The smoke, however, is prevented by six steam jets which impinge

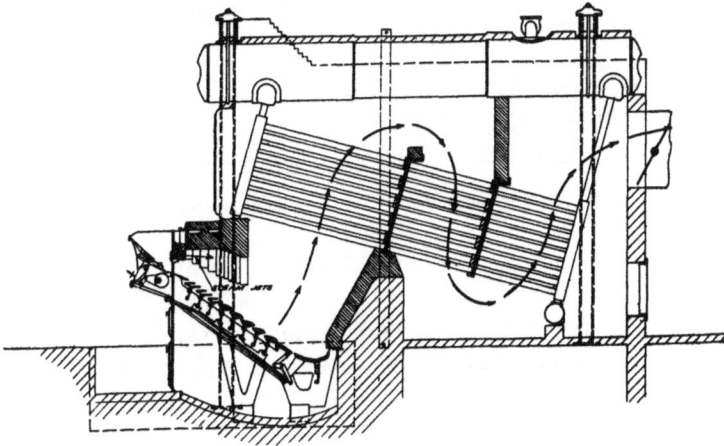

FIG. 56.—Babcock & Wilcox Vertical-pass boiler—Roney stoker

on the gases as they leave the fire and form a complete mixture with the oxygen of the air. The amount of steam required for the operation of these jets is about 5% of the total output of the boiler. While the smoke is prevented in this way, the method is an inefficient one and should only be used where the other forms of baffling shown herewith are impracticable.

Fig. 57 shows the Burke furnace in connection with the Edgemoor boiler. As will be seen from the illustration, the combustion is accomplished not only by the assistance of the coking or ignition arch in the furnace, but also by means of the lower baffling of the boiler. This arrangement of furnace

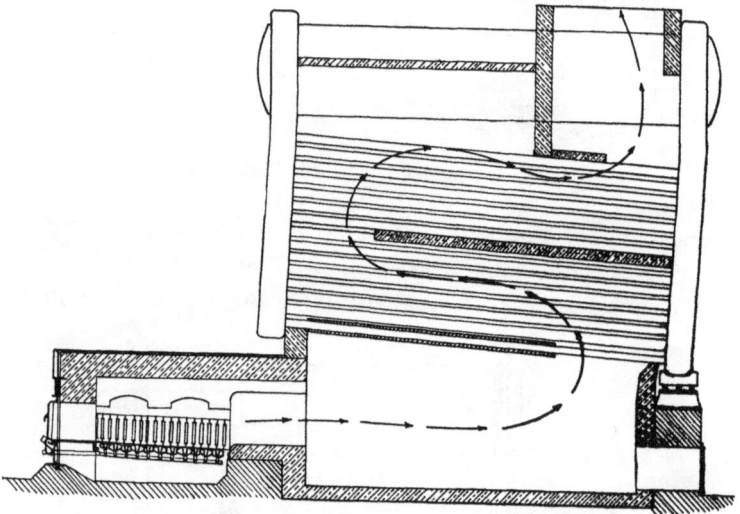

FIG. 57.—Edgemoor Horizontal-pass boiler—Burke furnace

and baffling is used in a district heating-plant and has been found very satisfactory.

Fig. 58 illustrates the combination of chain-grate furnace with a Sederholm boiler. The gases as they leave the chain grate impinge on the coking arch and the combustion is completed in the combustion space before the gases reach the water tubes of the lower part of the boiler. As shown in the illustration the upper portion of the boiler is the ordinary fire-tube type, the gases passing through tubes inside the boiler.

Fig. 59 shows a combination of the Murphy type of furnace with a Geary boiler. Complete combustion and consequently smokeless operation is made possible by the lower horizontal

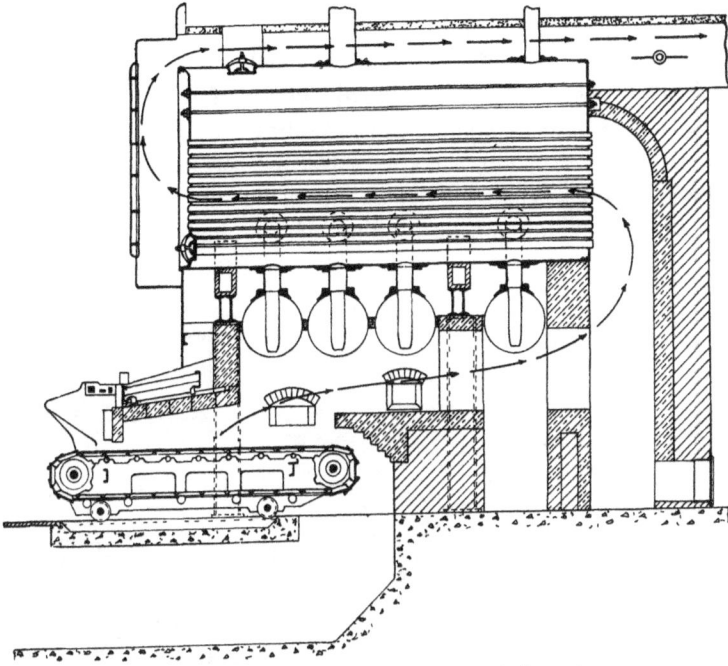

Fig. 58.—Sederholm boiler and Green chain grate

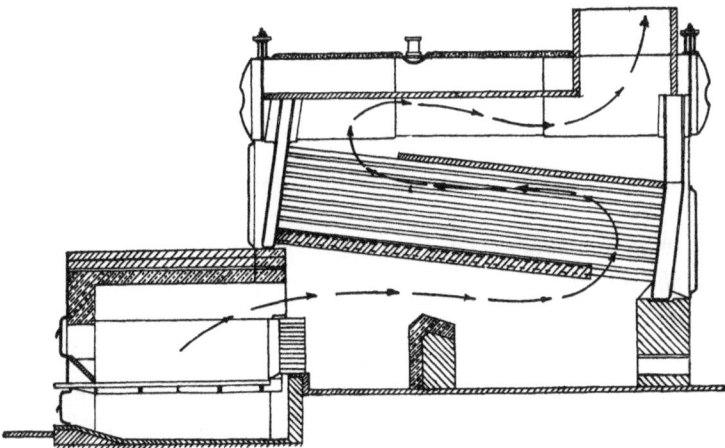

Fig. 59.—Geary Horizontal-pass boiler—Murphy stoker

baffling of the boiler, which is directly above the combustion chamber.

Fig. 60 shows a combination of a chain grate with Babcock & Wilcox boiler having combined vertical and horizontal baffling and is used in a large office building recently erected. As shown in the diagram, the complete combustion of the gases is effected by the coking arch above the chain grate, and also by the lower horizontal baffling of the boiler. This plant

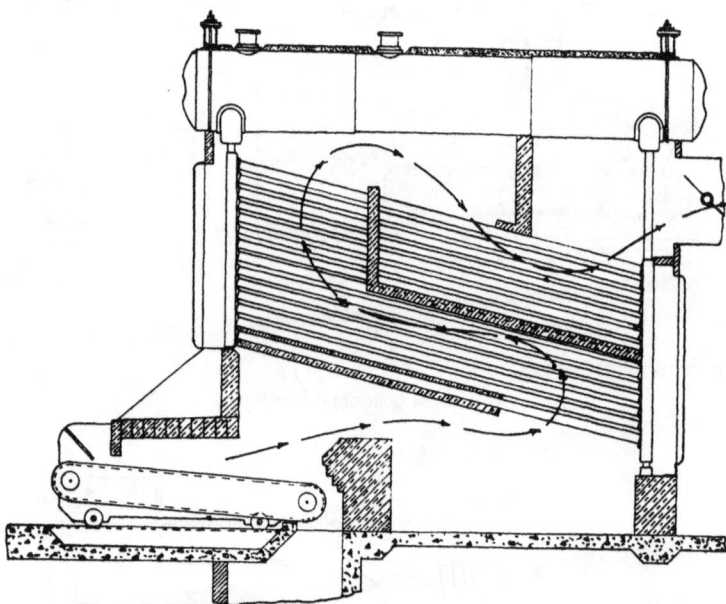

Fig. 60.—Babcock & Wilcox Combination-pass boiler—Chain grate

although it has operated for only one year has been found to operate through a wide range of conditions without smoke.

Fig. 61 shows a combination of Babcock & Wilcox boilers with chain grates in a very large representative power plant in the middle West, the arrangement consisting of a combination of horizontal and vertical baffles. As will be seen in the diagram the arrangement is the reverse of the ordinary Babcock & Wilcox boiler settings, the front of the boiler being

lower than the rear. This enables a greater travel of gases
before they impinge on the cold tubes.

The foregoing are simply typical forms of furnaces and
boiler arrangements and will perhaps be sufficient to illustrate

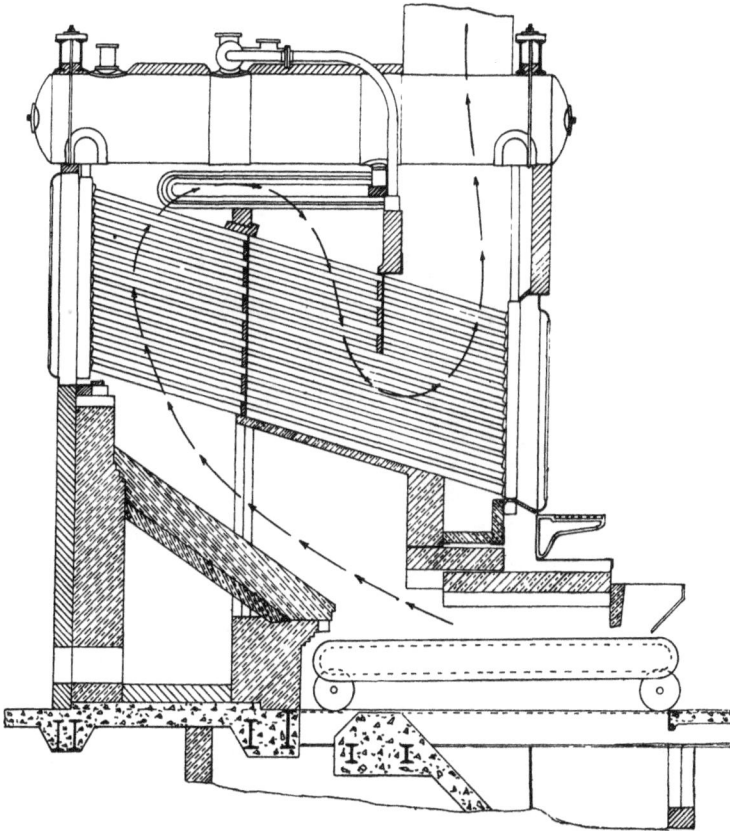

FIG. 61.—Babcock & Wilcox Vertical-pass boiler and chain-grate stoker

the different methods that are used to obtain complete combustion and therefore smokeless and efficient operation.

Typical Hot-water Heating Plant. Figs. 62 and 63 illustrate
the supply station for the large hot-water heating system referred
to in Chapter III. The boiler equipment consists of three 500-
horse-power water-tube boilers with chain-grate stokers. These

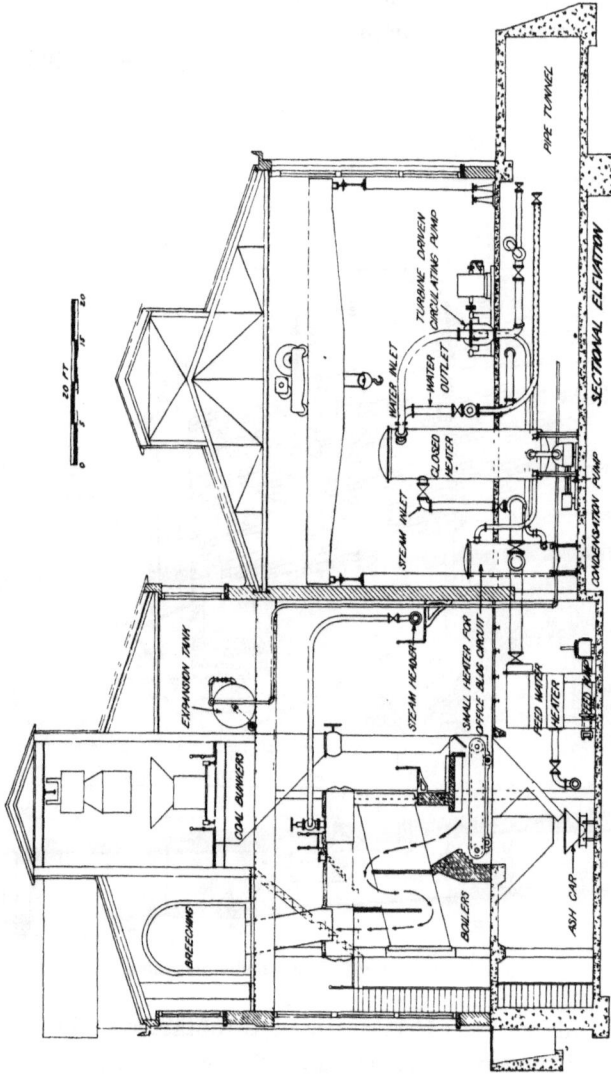

FIG. 62.—Supply station for the Hot-water heating-system in plant of Crane Company, Chicago

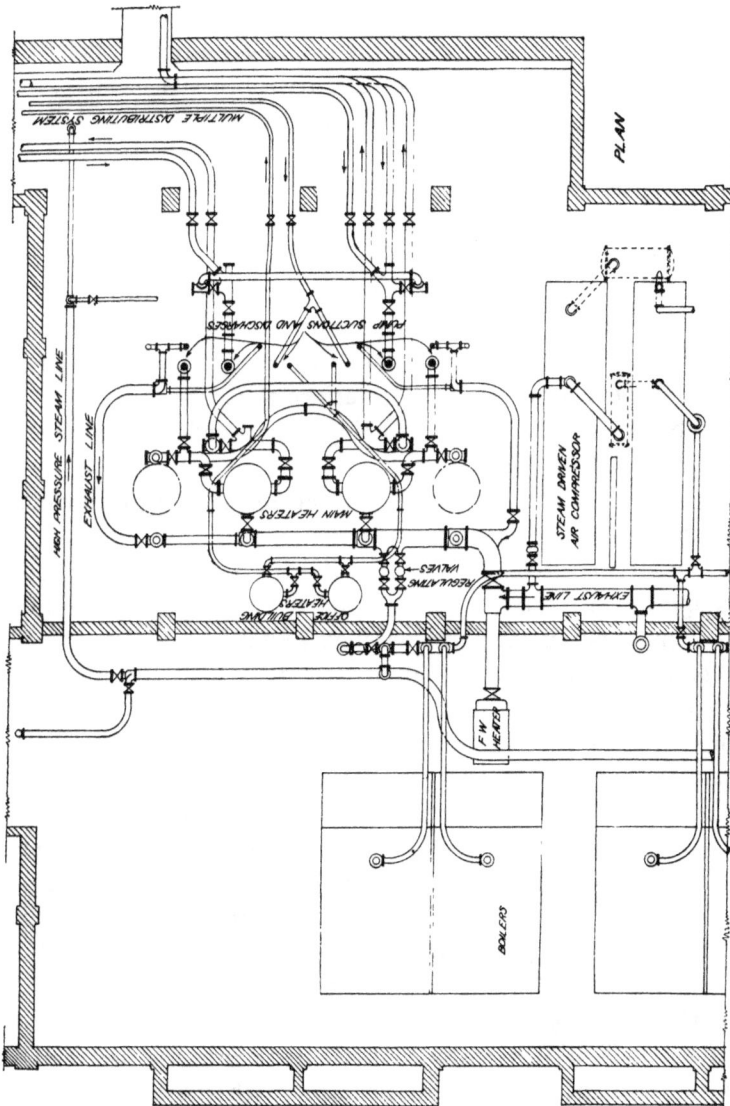

FIG. 63.—Supply station for the Hot-water heating-system in plant of Crane Company, Chicago

are served by a very simple but complete system of coal and ash-handling machinery, eliminating all belt conveyors and substituting a telpher system for transferring coal from the switch track to any desired location in the main overhead coal bunker. The ashes are handled in small cars on a basement track and hoisted by an elevator to the ground floor level where they are dumped and used for filling.

The steam is distributed through a basement header and supplies two large air compressors, boiler-feed pumps, vacuum pumps and turbine-circulating pumps. The exhaust steam from these units, supplemented by the required amount of live steam introduced through a regulating valve, enters the water heaters through which the circulating water passes. These heaters have a combined heating surface of 4050 sq.ft., separate units being provided for the administration building and shop buildings. These units are installed in duplicate and space has been reserved for future extensions and installations of double the present capacity.

The steam turbine driven centrifugal pumps, also installed in duplicate, are rated at 300 gal. and 3500 gal. per minute for the administration and shop buildings respectively. The pump discharge pressure averages 50 lbs. per square inch and the suction pressure 13 lbs. per square inch. The outgoing water temperature is designed for about 170°, with a 30° drop. The expansion tank is located some 40 feet above the pump suction and connected to same by a 3-in. line.

The present system comprises two complete circuits of two 9-in. lines each and one circuit of two 6-in. lines. The ultimate load will require two additional 9-in. circuits. The quantity of water circulated per square foot of radiation is approximately 6 to 8 lbs. per hour and the velocity through the pipe-lines from 5 to 7 ft. per second.

Each building service is equipped with pressure and temperature control devices. All radiation with the exception of the offices is made of pipe-coil.

This plant supplies the heating service for a very large manufacturing establishment which, with the exception of the air compressors and pumps above mentioned, uses central station service for its entire lighting and power requirements, aggregating several thousand horse-power.

The success attained in many instances with hot-water systems would seem to indicate that in the future they will increase in popularity with industrial plants, where the load is comparatively compact and the distributing lines available for repair.

Average Efficiency of Boilers. The average efficiency of various types of boilers is a matter which is always open to discussion. Engineers frequently claim an evaporation of 8 or 9 lbs. of steam per pound of coal and these efficiencies are backed up by reports of elaborate tests. The efficiency of a boiler under test conditions, when operating at full load with the particular kind of coal required, and when the fireman is under the direct eye of the testing organization is one thing; the average efficiency that can be secured from the average plant is altogether another matter. This net efficiency can be determined only by installing accurate instruments which measure the net output of steam as compared with the total quantity of coal burned. The net efficiency includes all the steam losses on account of blowing down of boilers, as well as losses due to mistakes of firemen, to improper coal, and to the various emergencies which arise in actual practice. It is the figure which must be used in figuring the actual cost of the production of steam.

It is probably fair to state that in a well-managed boiler plant the evaporation will run from $5\frac{1}{2}$ to 6 lbs. of steam per pound of coal where ordinary types of Illinois bituminous coals are used. The Eastern coals will doubtless give a somewhat higher evaporation but even with these coals the high claims sometimes heard for efficiency should be scrutinized very closely. The average net efficiency of a boiler takes into account its all-day performance including both the night and day load. Frequently at night the boilers are producing little or no steam and operating under banked fires. There are days in the winter time when the coal is thick with snow and ice and other times when the washers at the mines are not operating properly and it is impossible to obtain just the right quality of coal. All these drawbacks and disadvantages are included in the net result that is obtained. Of course, the object in plant operation is to reduce these losses to a minimum, but no matter how carefully the plant is operated there are inevitably certain losses which are likely to be ignored in the theoretical calculation.

Importance of a Thorough Knowledge of Plant Conditions. It has not been deemed advisable in this volume to enter upon the strictly technical phases of station operation, since those questions have been exhaustively treated in various works published within recent years.

Engineers have long striven with every power at their command to improve the economy of prime movers—engines and turbines—and at the present day the limit of practical efficiency has been very nearly attained. This concentration of attention to the steam-consuming units has resulted in corresponding laxity in attention to details pertaining to steam production. Possibly the more congenial atmosphere of a well-kept engine room leads to a natural aversion for the dirty and grimy boiler room. Be that as it may, the fact remains that it is in the steam-generating equipment, that the most important future improvements may be looked for.

The boiler room of a district heating company should be considered in the light of a large laboratory, where the chemical combustion process is carried on according to scientific principles. A study of the illustrations showing different types of furnaces, boilers and baffling would suggest that a great deal may be gained by investigating the performance of ·each type of unit with reference to:

1. The efficiency of combustion.

2. The efficiency of heat transferrence from the gases to the water within the boiler.

3. The efficiency in relation to capacity or rate of driving the boiler.

There is no reason why every large boiler plant should not be equipped with the comparatively simple and inexpensive testing instruments for gas-analysis and for taking drafts and temperatures. Their use is absolutely imperative if the best results are to be attained. In many plants operating with supposed economy, the data secured by the aid of these testing instruments when properly interpreted disclose unsuspected and preventable losses.

A gas analysis instrument for determining the percentage of the constituents in the flue-gases, carbon-dioxide, carbon-monoxide and oxygen, when used in connection with the draft gauge indicates the completeness of combustion and warns the fireman of undue excess of air supply, enabling him to control the operation by damper adjustment very easily.

The draft readings taken through different parts of the boiler setting from furnace to damper indicate the relative drop in draft and if their significance is understood may lead to a rearrangement of the baffling between the tubes so that greater effectiveness will result. The flue-gas temperature is largely controlled by draft adjustment and by properly proportioning the different parts of the heating surface.

The present tendency favors higher rates of evaporation and overloads on the boilers. Some companies are probably deterred from extending their business because of a mistaken hesitancy over operating their boilers at capacities much in excess of their rating.

Among the works which could be very profitably consulted for more specific information along these lines, may be mentioned:

Transactions—National District Heating Association, 1914, U. S. Bureau of Mines—Bulletins 8, 18, 39, 40 and 49.

Gebhardt—" Steam Power Plant Engineering."

Jos. W. Hays—" How to Build up Furnace Efficiency."

Gill—" Gas and Fuel Analysis for Engineers."

In closing this chapter, it may be well to speak a word of caution against the danger of jumping at conclusions. Every plant operator frequently meets with fluent salesmen who have some device which is supposed to cut down the cost of plant operation to a very great extent. Undoubtedly some of these devices are of value, but before making expensive changes in plant conditions, it is well to study the subject in all its bearings.

Years ago, one of the greatest educators in England while visiting a college was asked to make a few remarks to a class in geology. His entire speech consisted of the following words:

"Gentlemen: I wish to warn you against the danger of making conclusions from insufficient data."

Perhaps there is no field of knowledge in which this advice is more applicable than that of steam plant operation. Many plant owners lose large sums of money through lack of accurate information. It would be well to take a lesson from Pasteur, the great French scientist, who never accepted a theory unless it was confirmed by the most careful and painstaking tests. Even the most careful tests should be checked to see that the results agree.

CHAPTER VI

METHODS OF ESTIMATING HEATING REQUIREMENTS IN BUILDINGS

The first question that arises when a prospective customer is approached by the solicitor from the heating company is " What will the service cost? " The answer to this query is not always immediately forthcoming, and, indeed, when the many elements are taken into consideration, the reason for this is apparent.

Large Buildings. Among engineers the question of heating large buildings has caused much disagreement and discussion. This is due in a large measure to the lack of data covering actual operation and to the fact that there has been a tendency to subordinate its importance when considered in relation to power and lighting supply, the disposition being to consider it merely as a by-product of the isolated plant. It is remarkable what a very small amount of authentic information is available in comparison with the volume of records pertaining to electrical requirements.

Many investigators have at different times proposed methods of estimating the heating requirements of buildings and have reduced their calculations, presumably checked by observations of actual results, to formulæ. The attempt is usually made to combine the many variables into a simple expression which may be safely employed by those not thoroughly familiar with this branch of engineering. These efforts have met with more or less success as will appear later.

CLIMATE

Fig. 64, based on data available by Weather Bureau records, shows the temperature variations and the modifying influences

in the United States. This chart gives the isothermal lines
connecting points having equal degrees of normal annual temperature, and emphasizes the important effects of the entire
Pacific coast in increasing the temperature. The influence of
the high mountainous region of the west is indicated by the
southward dip of these isotherms showing much lower temperatures in this locality. In the east central portion of the country
contiguous to the Great Lakes the cold water retards the normal
spring temperature rise, this feature not being indicated in the

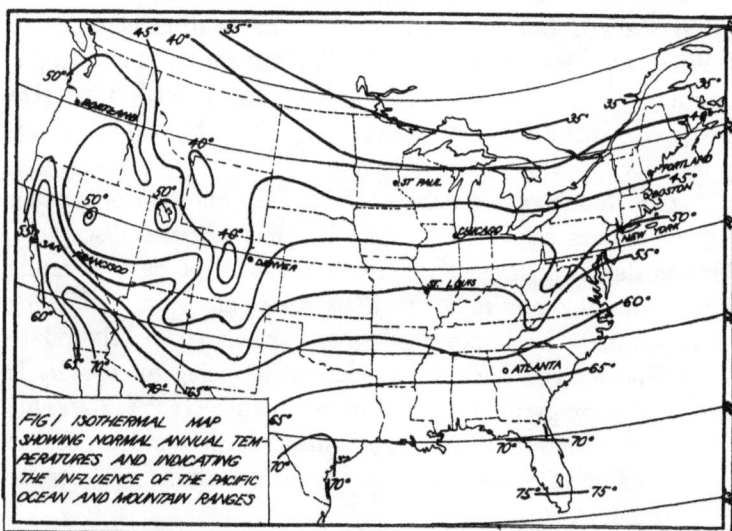

Fig. 64.—Isothermal Map showing normal annual temperatures and indicating
the influence of the Pacific Ocean and mountain ranges. (Cox and Armington).

chart.   Table I presents average monthly temperatures for
many cities based on long periods of observations.

The term " temperature " refers to the sensible heat of the
earth's atmosphere based upon a single observation of the
thermometer. General conditions of weather, however, enforce
the use of the terms "mean " or " average " temperatures for
expressing the average values. The method of arriving at the
monthly or yearly temperatures is to average the successive
daily observations during the corresponding period, but the
general custom for daily temperatures is to strike a mean between

the maximum and minimum observation. This has been found to be justified as it results in negligible error. So-called " normal " temperatures are averages of long periods or series of years from which individual years deviate to a greater or less extent. The use of normal temperatures in computations of heating requirements is essential.

There is considerable uncertainty in comparing meteorological data in various cities since so much depends upon the construction of and methods used at different observatories. It is sometimes said that temperatures at the earth's surface are slowly, but steadily increasing each year, allowing of course for occasional reversions. This statement sounds plausible but is not substantiated by records. Careful students of the subject state that since the beginning of history there has been probably very slight deviation from the normals observed during the last century, both in this country and Europe. There is, however, an explanation of this apparent contradiction, and one which concerns heating companies to a large extent.

The records reveal the fact that the general temperature range is less in large cities than in surrounding open country. The reason for this is to be found in the different environment offered by the city.

" 1. Temperatures in open areas of country are lower at night because of increased radiation of heat, the atmosphere of the average city being charged with smoke, dust and vapors which serve as a blanket to prevent radiation.

" 2. The immense amount of fuel consumed with the accompanying discharge of heated gases operates to increase the sensible temperature of the envelope of air in the vicinity of cities, at the same time increasing the moisture absorbing capacity of the air thereby adding to the fog or mist.

" 3. A considerable amount of heat is radiated from the large buildings in the congested districts."

The above remarks illustrate the need of care in noting the conditions under which temperature records are obtained. There is a wide variation in elevation and exposure between different observation stations.

In Chicago, for example, temperatures in the suburbs often show much lower values than in the city. This difference has been found to be as high as 13° and accounts for the fact that snow and ice may be encountered outside the city when the Weather Bureau record is well above freezing temperature. Zero temperatures are quite rare in coast cities and the rigors of winter are much tempered by the influences of the ocean currents. The Great Lakes serve to break up both hot and cold waves since, when the wind blows over a considerable expanse of water, there is an interchange of heat due to the fact that water has a much higher specific heat than air and also is much slower to receive or impart its heat by conduction.

The temperature changes being largely due to radiation, it will be seen that cloudiness from whatever cause serves as a blanket which interferes not only with the heat received from the sun in daytime, but also prevents the normal radiation and loss of heat during the night. Therefore, clear atmosphere, with little rain is conducive to much greater daily changes in temperature so that, with other conditions equal, the temperature varies as the hours of sunshine; and relative humidity is affected through temperature changes resulting from the same cause.

The various factors entering into the problem of heating buildings may be summed up as follows:

1. Duration of the heating season.

2. Temperatures during the above period and thermal difference between outdoors and indoors.

3. Modifying influence of wind velocity, humidity and related effects.

4. Hours of heating required due to daily variation of temperature.

5. Type, construction and exposure of building.

6. Method of heating—direct, direct-indirect, and indirect systems and amount of ventilation.

7. Methods of regulating temperature.

8. Class of service, varying with the purpose for which the building is used and the peculiarities of tenants.

9. The type and efficiency of the steam-generating equipment in central-heating plants and individual buildings.

10. The economy of heat production involving fuel, labor and miscellaneous charges.

It is obvious that all the above variables have their essential part in the derivation of any formulæ. It is of course desirable to combine these factors into a simple form of expression which will be applicable to as great a variety of conditions and different localities as is feasible. To this end a discussion of the relative values of the components entering into the problem is in order.

(1) and (2), **Duration of Heating Season and Temperatures.** Perhaps as satisfactory a method as any for determining the duration of a heating season is one proposed by Mr. E. F. Tweedy and published for the first time as part of the " Report of the Committee on Steam Heating," National Electric Light Association, 1912. This method is graphical and is presented in a considerably modified form as Fig. 65. As a basis for these curves, it is assumed that:

" (a) The temperature maintained within the building is 70° F.

" (b) Artificial heating is required whenever the mean daily temperature falls to 60° F., this seeming discrepancy resulting from the daily variation in temperature. This is apparent from Fig. 66, which shows that the temperature is considerably higher during the daylight hours when heat is required than during the night. The low nightly temperatures serve to reduce the mean average daily and monthly values and for this reason the reports of the Weather Bureau are likely to be misleading when used for exact heating calculations. The precise temperature below which heating is required can only be approximated, but the value given probably represents the average condition throughout the country."

The curves are obtained by plotting the observed mean monthly temperatures of various cities in many localities as given by Weather Bureau reports. Where these curves intersect the line marked 60°, it is assumed that the horizontal dis-

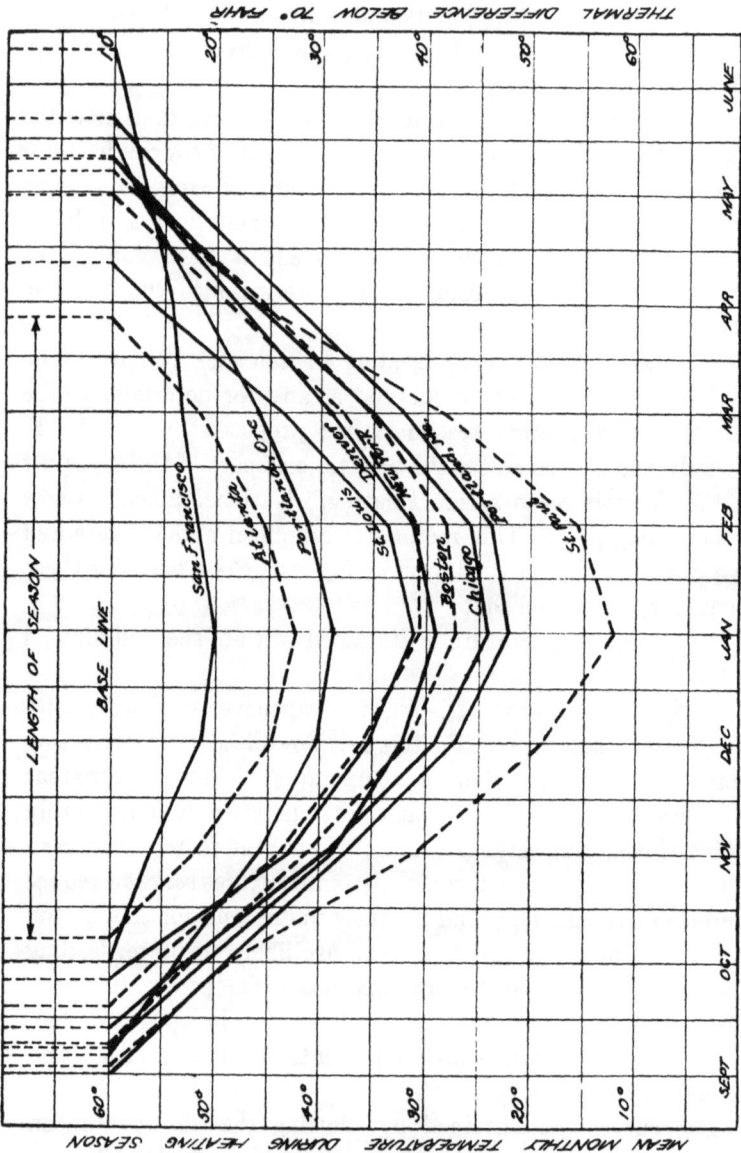

FIG. 65.—Diagram showing approximate heating requirements in different cities

tance included between the intercepts represents the number of days that heating will be necessary and that the areas included between the 60° base line and the curves will be a measure of the relative heating requirements. The data from which Fig.

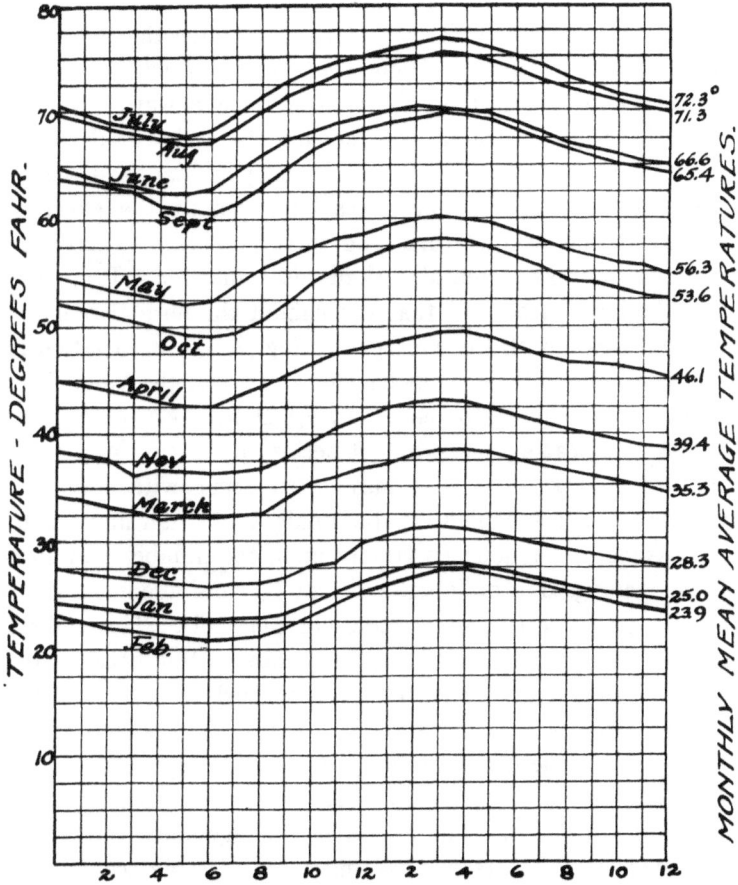

FIG. 66.—Diagram showing hourly variations of temperature in Chicago, Ill.

65 is derived are given in Table I. Table II presents the deductions from these data.

(3) **Modifying Influence of Wind Velocity, Humidity and Related Effects.** The effects of wind velocity and humidity are generally lost sight of in heating calculations, but it cannot be

doubted that many of the peculiar results obtained in different parts of the country having approximately equal latitude may be ascribed in part to these influences.

The velocity of air movement determines largely the amount of air leakage in buildings due to infiltration on account of velocity pressure, as well as heat loss by convection from the wall and glass surface.   High wind velocities will cause an increased leakage into the building from the windward side, and, as has been demonstrated, will increase the withdrawal of air by the suction effect on the lee side, greatly adding to the ventilation losses.   Repeated instances have proven the effect of winds with low temperatures and these observations have led to more or less experimental work in this connection.   It appears from the best authorities that throughout the normal range of temperatures during the average heating season, a variation in velocity of one mile per hour has approximately the same effect as one degree change in temperature.   It has been noted that the extreme heating loads do not necessarily occur on days of minimum temperature, but may often occur with moderate temperature accompanied by abnormally high wind movement.

Leakage of air depends largely upon the number and sizes of the openings in a building, and may be considered as a function of the total area of openings in the walls.   This follows from the fact that there is an approximate relation between the area and the perimeter of the average-sized window opening. The perimeter which is assumed to include the length of the center cross-frame in the ordinary two-section movable window is the only available means of direct access of air assuming an otherwise leak-proof building.   This leakage may be largely prevented by weatherstrips making a considerable economy in fuel.

The direction of wind movement has a bearing upon the setting of radiation and explains the prevalent practice of allowing arbitrary increases in the amount of radiating surface on the most exposed sides of the building.

Generally speaking, the hourly curve of wind velocity follows the temperature curve quite closely, this being occasioned by

Table I, MONTHLY AND ANNUAL NORMAL TEMPERATURES FOR VARIOUS CITIES IN DEGREES FAHR.

	Portland Oregon	St. Louis Missouri	St Paul Minn.	Atlanta Georgia	Boston Mass.	Portland Maine.	Chicago Illinois.	New York N.Y.	San Francisco, Cal.	Denver Colo.
January	38.7	31.0	11.9	42.2	27.0	22.0	24.1	30.6	50.0	29.0
February	41.4	33.5	15.4	45.2	28.0	23.8	25.7	30.7	51.6	31.0
March	46.1	43.5	28.2	52.4	35.0	32.0	35.2	37.8	53.2	39.0
April	51.8	56.1	45.8	61.1	45.3	43.0	46.4	48.7	54.1	48.0
May	57.7	66.5	57.7	69.5	56.3	53.5	56.7	59.8	56.5	57.0
June	62.5	75.1	67.2	75.6	65.8	62.6	66.7	69.0	59.0	66.0
July	67.3	79.1	72.0	77.6	71.3	68.0	72.6	74.1	59.0	72.0
August	66.6	77.2	69.7	76.1	68.9	66.2	71.4	72.6	59.0	70.0
September	61.3	70.0	60.5	72.1	62.7	59.6	64.9	66.5	61.0	63.0
October	53.7	58.4	48.4	62.4	52.3	49.1	53.1	55.6	60.0	51.0
November	45.7	43.4	31.0	51.9	41.2	37.6	39.5	43.9	56.2	39.0
December	40.2	35.5	18.8	44.6	31.6	27.1	29.5	36.0	51.4	32.0
Mean Yearly Average	52.8	55.8	43.9	60.9	48.8	45.4	48.9	52.1	55.9	50.0

the warming up of the ground which induces the air currents to ascend more rapidly. Hence, it follows that generally the highest velocities occur during the warmest hours. However, it will be found in considering the monthly averages that other influences are at work and that the coldest months as a rule are attended with highest velocities.

" Humidity " is a term employed in referring to the moisture content of the air, such moisture being in the invisible state. Dry air contains but a very slight quantity of moisture and when a relatively large amount is present the air is said to be humid. " Absolute humidity " is the actual quantity of moisture in a given quantity of air, whereas " relative humidity " is the percentage of the maximum capacity of the air to contain moisture, or its per cent of possible saturation. Rise in temperature of course increases the capacity of air to absorb vapor. The subject of relative humidity has not received the attention it deserves, and it is not generally recognized what a vital effect it has upon the health and comfort of the earth's population.

The human body is at all times throwing off heat and moisture from the skin and lungs which is absorbed by the air. If the air is already saturated with moisture the process of transfer is interfered with and the result is discomfort. On the other hand, very dry air in contact with the skin will accelerate the evaporation and cause quick absorption of perspiration and cooling. The relative humidity for a given temperature of outside air is greatly reduced by any increases above that temperature, thus 50% relative humidity at 50° falls to 25% at 70°. This emphasizes the fact that the humidity of the air inside a building may decrease at an alarming rate unless some artificial method is employed for adding moisture. With the modern methods of humidifying air in ventilating equipment, it is possible to exercise definite control over the relative humidity and a substantial saving in heat may result, since if the proper relative humidity is maintained it is possible to keep the temperature somewhat lower. This saving is offset somewhat by the cost of operating the equipment. There has appeared on the market recently a device which is very effective in maintaining the proper amount of moisture within the room. This device is fitted on to the radiator in the same manner as an ordinary air-valve and allows the escape of a portion of the steam which is absorbed by the dry air, the operation being entirely noiseless. While no authentic tests are available, it is reasonable to believe that this method results in a fuel saving as well as being instrumental in adding to the comfort of the occupants. The leading authorities agree that a relative humidity of approximately 55% produces best results.

In certain parts of Europe, notably in England, it is well known that 60° is considered the standard indoor temperature and this may be due in part to the greater relative humidity. It is probable that these differences explain somewhat the variations in heating requirements between sea-coast cities, where the relative humidity may be higher, and inland localities; but as no authentic data can be cited in support of this, this factor of humidity cannot be assigned any absolute value in calculations.

(4) **Hours of Heating Required per Day.** Table II indicates the duration of the heating season in days but does not consider the question of partial, continuous or intermittent service during that period. This point is of importance in heating calculations, although like a great many other factors it is somewhat difficult to assign definite values, due to the lack of data. At first thought, it seems a reasonable assumption that if the steam be turned off a building during the night, a fuel saving should result. Since the outdoor temperatures are considerably lower during the night, the heat loss would be increased, other conditions being equal, although it is probable that this would be modified by the lower wind velocity and other contributing causes. Except in special cases there seems to be no reason why normal temperature should be maintained at all times unless it is found that the building loses heat with such rapidity that it is difficult to heat the premises up again to the required temperature.

If sufficient radiation is provided the building should be easily brought up to temperature before being occupied in the

TABLE II.

SHOWING THE LENGTH OF HEATING SEASON, MEAN AVERAGE TEMPERATURE DURING HEATING SEASON, THERMAL DIFFERENCE AND RELATIVE EXTENT OF HEATING REQUIREMENTS BASED ON TEMPERATURE ONLY.

Location.	Latitude.		Days.	Average Temperature during Heating Season.	70° minus Average Temperature during Season.	Per cent Requirements based on St. Paul Conditions.
	Deg.	Min.		Deg. F.	Deg. F.	
St. Paul, Minn . . .	44	0	248	33 5	36 5	100 0
Portland, Maine. . .	43	40	264	40 7	29 3	85 6
Chicago, Ill	42	0	243	39 2	30 8	82 75
Boston, Mass.	42	20	247	40 6	29 4	80 4
Denver, Colo. . .	39	50	248	41 5	28 5	78 1
New York, N. Y	40	40	223	42 0	28 0	69 1
Portland, Ore .	45	40	248	47 9	22 1	60 5
St. Louis, Mo	38	31	197	42 6	27 4	59 65
San Francisco, Cal	37	40	251	54 4	15.6	43 3
Atlanta, Ga	33	40	171	49 4	20 6	39 0

morning. This whole question is open to debate and engineers are far from being agreed upon the matter. In extreme cold weather it is unwise to leave a building unheated since the indoor temperature may fall below freezing and occasion much damage, especially if a sprinkler system is installed. At just what temperature it is permissible to reduce the hours of service is a question for experiment, but in general, it is safe to say that above 15° to 25° outdoor temperature it is advisable to reduce the supply of heat during the night. Unless some automatic control is possible where the building is closed up for the night, the temperature is very likely to rise up to 80° or above, thus increasing the transmission of heat. The greatest opportunity for economy along this line is in industrial or manufacturing buildings. Next will come stores, department stores and office buildings. In hotels and similar buildings the demand for heat is so continuous that very little opportunity is offered for shutting off the heat at night. Generally speaking, department stores require heat during severe weather for from 12 to 18 hours, office buildings, 16 to 24 hours and hotels continuously. For buildings buying heating service by meter a direct incentive is furnished the consumer to economize in every possible manner, but co-operation by the heating company is required to assist in locating preventable wastes of the heating service.

(5) **Type, Construction and Exposure of Building.** Large modern buildings are generally constructed with a skeleton, or framework of steel columns and girders, enclosed by a brick wall and finished on the outside with brick or terra-cotta tile. With such tall buildings it is necessary to use the lightest material available in order to decrease the weight on the steel work and foundations. In doing this, of course, the thinness of the walls becomes of importance from the standpoint of heating calculations. Such buildings have little capacity for storing or retaining heat, which is in contrast to what is found in buildings constructed of massive masonry. In the former case—the modern building—heat must be furnished for a much longer daily period than in the latter, due to the more rapid cooling effect. Furthermore, the modern structure is designed with a view to utilizing as much of the exterior as possible for window space as by so doing the lighting conditions are vastly improved. In fact, some buildings are practically 40% to 45% glass area, and the heat loss from such buildings is proportionally high. The average modern building shows a ratio (glass to total exposure) of approximately 30%.

One peculiarity which has sometimes been noted in very high structures is the draft effect due to the inrush of cold air through openings in the lower floors. This air upon being heated rises rapidly through the various elevator and ventilating shafts and causes a partial vacuum effect with a useless expenditure for heating the large volume of air. Some large buildings recently erected are so designed as to offer very little resistance to wind effects. In estimating the fuel consumption for these buildings it is necessary to increase the ordinary requirements 10% to 20% for this reason alone, as experience has proven the difficulty of overcoming the influence of the cross-currents of air. The location of a building with reference to the direction of exposure and the proximity of the surrounding structures must

be carefully considered since this largely affects the heat loss
due to wind velocity.

(6) **Method of Heating.** In Chapter III, pp. 96 to 104, the
various methods of arranging the piping systems within
buildings are set forth. This classification shows that three
methods are in use for heating—(a) direct, (b) direct-indirect,
and (c) indirect.

Direct heating is the common practice for nearly all classes
of buildings since it is less costly both in installation and operation. Frequently the direct-indirect or semi-indirect is used,
but its advantages from the standpoint of ventilation are counter-
balanced by the expense of operation and difficulty of installing.
Indirect heating requires much greater steam consumption and
is the most expensive form of heating system.

During the past few years the subject of mechanical ventilation in buildings of all types has been receiving increased
study on the part of engineers. The passing by the different
states of compulsory ventilation laws has perhaps been largely
responsible for the recent study and investigation of the subject.
The heating of buildings by direct radiation has been worked
out on scientific lines and in a short time corresponding rules
for the solution of ventilation problems will have been formulated. The subject of direct heating is nearly always opposed
to that of ventilation, and if there is a great amount of ventilation, either natural or artificial, excess heat must be provided.

With forced ventilation, it is desirable in order to exercise
perfect control over the system, to have the building as free
from leakage as possible so as to prevent the flow of air into the
building from the outside, in portions of the building where it
is not needed for ventilation; and to prevent the warm
air from escaping from places where it should be retained for
its heating value. With the building practically sealed, so far
as window sashes and crevices are concerned, it is apparent
that perfect regulation of heat and air supply can be obtained.
As stated above, ventilation is to be had only by increased
expense for fuel and, consequently, from a purely economic
standpoint, forced, or fan ventilation should not be employed.

The question is whether the benefits of improved ventilation and the health of the occupants of the building will outweigh the increased cost of same. While mechanical ventilation is in the majority of cases to be desired in preference to ventilation through the agency of open windows, yet in the case of office buildings, for example, the space which would be occupied by the air-ducts is of too much value to admit of the use of this system. Therefore, it will be found that the fan system of heating in office buildings is confined to the basement of the building, the corridors, and the lower floors if occupied by banks or similar institutions. Speaking in general, up to this time the indirect or fan-blast system of heating seems to have found its largest field in the heating of factory buildings and workshops. In some cases manufacturers have found that the increased efficiency of their workmen has more than offset the increased cost of heating the buildings.

In all cases, fan-blast heating and ventilation should be considered as a problem entirely separate from the direct heating of the building. Experience has taught that it is best to install a sufficient amount of direct radiation to take care of all the heat losses from the building, allowing the air from forced blast equipment to enter at a temperature but slightly above the room temperature, instead of at 100° to 130°. When heated to this latter temperature, as must be the case in purely fan-blast heating-systems, the air is entirely too dry for comfort and must be admitted to the upper portion of the room in order to prevent blowing directly on the occupants. Also by this arrangement of separate heating systems, and fan-blast systems, the latter may be shut down during the night when there is no need for ventilation and direct radiation used to keep the building warm. Unfortunately, very little accurate data is available concerning the difference in cost of operating the two systems in any given building. However, it may be said conservatively that a system of heating a building by fan-blast will cost about twice as much to operate as a corresponding system of direct radiation.

The amount of heating surface required in the blast-system

is much less than where direct radiation is installed, and this is due entirely to the higher condensation efficiency of the heater coil as compared with direct radiation. The average heater-coil in a blast-system may condense as high as from 1.2 to 2.2 lbs. of steam per square foot per hour, depending on the temperature of steam and velocity of the air, whereas the maximum condensation in the ordinary cast-iron radiator is approximately 0.3 lb. per square foot per hour. In other words, the condensation per square foot of radiation in a blast-system is from four to seven times that of direct radiation.

(7) **Temperature Regulation.** The question of temperature regulation has occasioned much discussion at times. It is rightly supposed that the installation of a complete thermostatic control-system should result in an economical use of steam. It is, however, difficult to maintain these systems in good condition at all times. Many installations in past years have been rendered useless as a result of improper methods of installation, and lack of proper maintenance; but the more recent examples tend to support the idea that an automatic control-system is very desirable. Reference has been made to the latest forms of heating systems, namely, the vapor and atmospheric, in which systems the thermostatic control is omitted on each radiator, since the inlet-valve must be operated by hand in response to the temperature changes. While it is not advisable to present data showing the actual saving due to automatic control, yet it is safe to credit it with a substantial reduction in steam consumption, unless some other equally efficient method of operation by hand is used. It will rest largely with the estimator as to the allowance to make for this factor but a reasonable assumption would be approximately 10%.

(8) **Class of Service.** In relative requirements from the heating standpoint, buildings may be classed as follows, neglecting the residence and very small business buildings:

> Theatres,
> Manufacturing buildings.
> Stores and Department Stores,

Loft Buildings,
Office Buildings,
Clubs, Hotels and Apartments.

The variation in requirements of different types of buildings is found principally in the following:

(a) Number of hours it is necessary to keep heat on the building.

(b) The use of open windows for ventilation, which practice is common in many classes of buildings.

(c) The temperature it is necessary to maintain.

(d) The pressure required on the radiation.

Hotels and clubs are most extravagant in the use of heat, followed by office buildings and department stores in the order named. Hotels are in their nature merely enlargements of the apartment house idea and for this reason their heating requirements are relatively much more severe than other buildings. The guests in a hotel seldom trouble themselves to co-operate with the management to reduce the use of light or heat. The result is shown in the large coal consumption of the average hotel, as compared with other classes of buildings. It is safe to assume that the average hotel will require as much as 25 or 30 per cent more heat than other buildings, and this point cannot be too strongly emphasized.

Office or Loft buildings are most commonly met with in this line of work and will be treated in more detail as regards heating than other types.

The steam heating of department stores is not usually as extensive, comparatively speaking, as the other types of buildings we are considering. The customers being generally in motion and in street attire, do not feel the necessity of much artificial heat, and in fact it must be kept down to a minimum to preclude complaints. The clerks in these stores, however, are more or less inactive and confined to small areas behind counters affording little opportunity for physical exercise. The well-known fact that large congregations of people such as will be found in churches, auditoriums and theatres will, after a short period of time, begin to feel decidedly uncomfortable if the space is heated artificially, applies to a certain extent to the department store.

(9) **Efficiency of Steam Generating Equipment in Building.** It will be preferable not to go into the details of this important branch of the subject, which has already been treated at some length in Chapter V; but since the present discussion applies more particularly to the operation of individual plants in large buildings or where steam is not supplied from the street, it may be well to add some observations covering such methods of operation.

The problem of the boiler plant located in the heart of a large city presents angles which are not met with in the usual central heating plant operating in the smaller cities. On account of the abnormally high rental value of city property, it is often deemed essential to economize in space even at the sacrifice of economy in the operation of the boiler plant. It is therefore obvious that the best results are impossible of attainment in many of these makeshift plants.

A study of the conditions illustrated in Figs. 70–78, Chapter VIII, will demonstrate the great variation in load conditions imposed upon the boiler plant. This not only applies to daily operation but to monthly and yearly operation as well. At certain periods it is necessary to operate the boilers at heavy overloads and at other times the load drops much below the boiler rating. The losses due to banking fires, as a precaution in case of sudden fluctuations in load is one of the detrimental features which produce low efficiencies. Many other points might be enumerated to indicate the impossibility of securing high efficiencies.

(10) **Cost of Steam Production.** The engineer who wishes to arrive at the cost of operating a building is interested directly in the economy of steam production. The cost of generating steam may be divided into several items, namely: fuel, labor, water, ash removal, repairs and miscellaneous costs and the fixed or overhead charges.

Fig. 67 gives a number of curves showing the operating costs

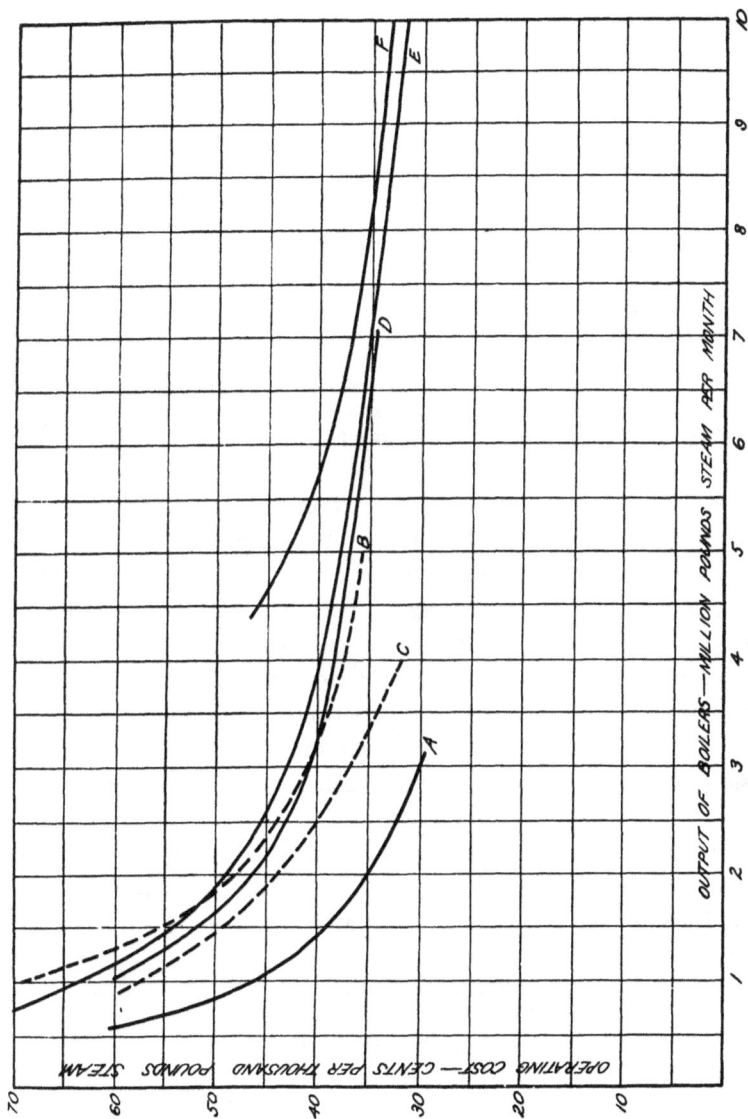

FIG. 67.—Curve showing cost of steam production

of producing steam in various plants located in Illinois, and indicates the wide range to be expected not only as between individual plants but also with reference to the total quantity of steam produced. The curves represent the operating charges, exclusive of overhead and supervision expenses in the following described plants.

Curve.	Type.	Boiler Equipment.	Furnace Equipment.
A	Medium sized hotel.	Return tubular.	Burke. Dutch Oven type.
B	Medium sized office building	Return tubular	Plain grate
C	Large club building	Scotch marine	Burke. Dutch Oven type
D	Large office building	Return tubular	Plain grate
E	Large office building	Water tube	Chain grate stoker
F	Department store	Water tube	Murphy stoker

Fig. 68 gives an analysis of the curves in Fig. 67, subdividing the costs into three main divisions: Coal, labor and sundries.

In each of the plants given the evaporation varies through considerable limits but averages about 5 to 6 lbs. of water per pound of coal. See Chapter V, p. 191.

In analyzing a plant the estimator must obtain as much information as possible regarding the condition of the equipment, the facilities for receiving and handling coal and removing ashes, the cost of labor in the particular locality and many other points too numerous to receive specific mention. The determination of the cost of steam is a matter which must be left largely to the judgment and experience of the estimator and no definite rules can be given to cover all conditions.

FORMULÆ FOR ESTIMATING HEATING REQUIREMENTS

The major portion of the preceding paragraphs presents information of a very general nature much of which is familiar, no doubt, to many readers. The main object in bringing out these points is to stimulate further investigation as to the relative weight of the various factors and if possible place the subject of heating upon a more scientific basis. It now remains

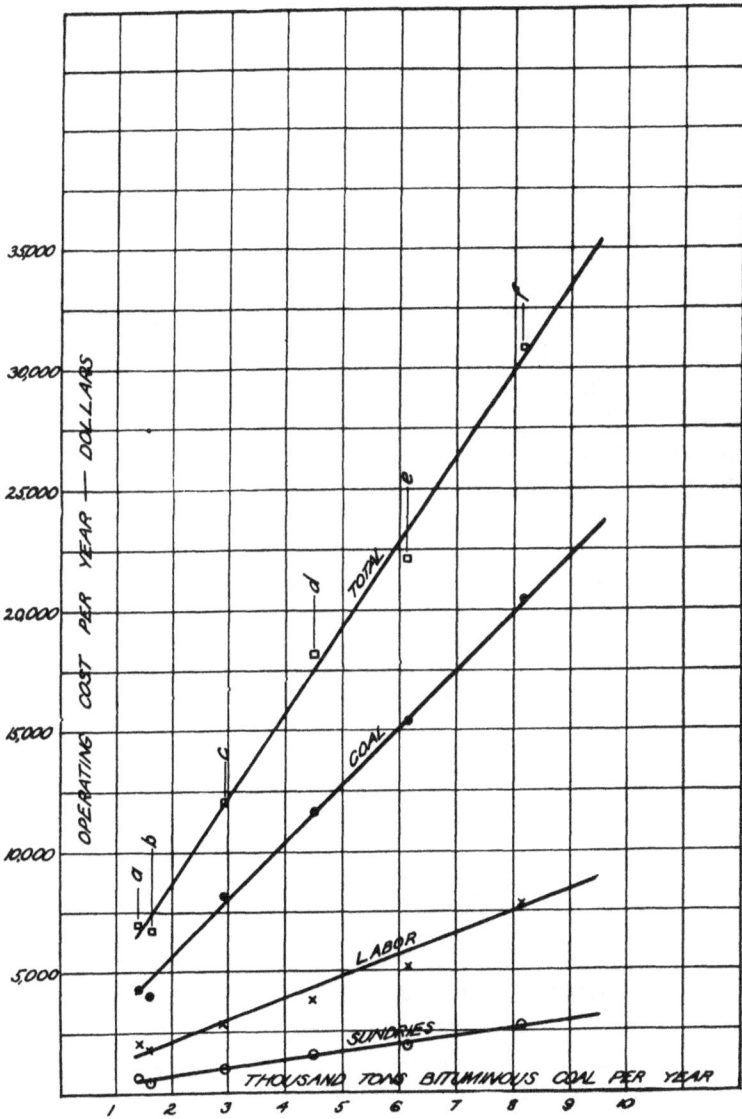

Fig. 68.—Diagram showing analysis of operating costs given in Fig. 67.

to collect such data as are at hand, with the idea of making practical use of same, and this is best accomplished by combining such of the above-mentioned variables as may be assigned somewhat definite values, into an arithmetical expression or formula. In the derivation of such a formula it would be preferable as far as possible to eliminate as many variables as would be consistent with reasonable accuracy. Otherwise its use would be limited to experts who are capable of selecting proper values for all such factors.

One such formula, credited to Mr. Davis S. Boyden, of the Edison Electric Illuminating Company of Boston, has proven suitable for determining the fuel requirements for any given building. On account of the many variables, however, it must be used with caution since one or more slight mistakes in selecting the values may cause a serious error. This formula appears decidedly formidable at first glance but if analyzed step by step will be better understood.

Tons of coal per year

$$= \frac{\left(\dfrac{V \times a}{60}\right) + (C_1 \times G) + (C_2 \times W)}{C_3 \times (130 - T)} \times L \times d \times h \times \left(\frac{34 \cdot 5}{e \times 2000}\right). -(1)$$

in which

V = Gross volume of heated space.

a = Average changes of air per hour.

G = Area of glass surface in square feet, plus 10% of the north and west exposures.

W = Net area of wall surface in square feet, plus 10% of the north and west exposures.

C_1 = Coefficient of heat transmission for glass = 1.0 B.T.U. per square foot per hour.

C_2 = Coefficient of heat transmission for walls, ranging from 0.2 for brick walls to 0.3 for stone walls.

C_3 = A coefficient varying with local conditions,

(Boston, Mass. = 5.4)

(New York = 5.7)

T = A factor covering transmission losses in pipe lines.

L = A factor allowing for any portion of the building not heated and also for the temperatures below 70°.

e = Pounds of steam evaporated per pound coal.

d = Duration of heating season in days.

h = Average number of hours of heating per day ranging as follows:

> Hotels—15 to 16 hours.
>
> Apartment houses—12 to 14 hours.
>
> Department stores and Office Buildings—10 to 12 hours.
>
> Theatres—6 to 10 hours.

In the above formula, several terms require explanation. The first term represents the heat required to raise the temperature of the air which enters through infiltration or is supplied directly for ventilation. It is based on the assumption that 55 cubic feet of air may be heated 1° by 1 B.T.U. Allowing for the volume occupied by partitions this is reduced slightly and the expression becomes:

$$\frac{\text{Gross volume} \times \text{air changes per hour}}{60} = \frac{V \times a}{60}.$$

The coefficient C_3 is determined by experiment and when divided into the heat loss will give the proper amount of radiation to be installed. A further assumption is made that each boiler horse-power will supply 130 sq.ft. of radiation. Considering the many variables in this formula it should be used with considerable care since its use is limited to engineers who are more or less familiar with its application.

A much simpler formula may be derived from the data given in the preceding pages. The following formula is based upon the experience of a heating company in Chicago and has been used successfully for more than ten years.

$$\text{Tons of coal per year} = \frac{S \times a \times b \times c}{e \times f \times 2000} = \frac{S \times K}{e}. \quad \ldots \ldots \quad (2)$$

where

S = Equivalent glass surface, or exposure, in square feet.

a = Coefficient of heat loss in British thermal units, per square foot of exposure per hour.

b = Thermal difference between the inside temperature and the mean average outdoor temperature during the heating season.

c = Number of hours in the heating season.

e = Pounds of steam evaporated per pound of coal under average conditions, throughout the entire year.

f = British thermal units contained in each pound of steam (assumed as 1000).

The quantity 2000 appearing in the denominator of the equation represents the number of pounds of coal per ton.

K = A constant obtained by assigning certain definite values for the quantities represented by a, b, c and f.

The values used to determine K, are dependent upon the variables outlined earlier in the chapter and must be ascertained by a study of conditions peculiar to the locality in which the heating company is operating. The outlines given in Sections 1 to 9 are for the purpose of assisting in the selection of the proper quantities. As a method of illustration, the formula used in Chicago will be worked out, and precisely similar operations will give the constant for cities in different latitudes.

DETERMINATION OF S (EXPOSURE)

The square feet of equivalent glass surface is computed by adding to the total area of the window openings one-tenth $(\frac{1}{10})$ of the net wall area, for walls of the usual fireproof brick or tile building of modern type. This relation is not strictly true when compared with values given by the different authorities who make the ratio of Glass : Wall = 10 : 2.5. However, the the former value is taken advisedly as will be seen later. Where there are double windows, the heat loss is cut in half. Skylights and sidewalk lights are considered as window area and the north and west exposures should be increased ten per cent. The method of computing S is shown in example, at the end of this chapter.

(a) COEFFICIENT OF HEAT LOSS

Authorities generally are agreed that the rate of heat transmission through glass with still-air conditions, is approximately 1.1 B.T.U. per square foot per degree per difference per hour. However, this coefficient must be increased in order to compensate for the leakage of air. In order to combine all the heat loss factors into one, the value (a) is taken equal to 2.0 B.T.U. from which it appears that approximately 50% of the heat loss may be ascribed to leakage and wind movement. This assumption does away with the necessity of calculating the number of air changes per hour, eliminating one additional factor from the formula.

(b) TEMPERATURE DIFFERENCE

This value is found by aid of Weather Bureau Records, which give the mean average temperature during the heating season. In Chicago, for which city this formula is being developed, this difference is represented by the difference between 70°, the temperature maintained inside the building, and 39°, the average outside temperature. This gives 31° as the value of (b.)

(c) NUMBER OF HOURS

This factor is determined also from Weather Bureau Records. For Chicago, it is taken as 243 days, or from October 1st to May 31st. Obviously, steam-heating is not required throughout the entire months of October and May, but if (c) is decreased the value for (b) will, necessarily be increased.

(e) EVAPORATION OF STEAM PER POUND OF COAL

This has been treated under "Type and Efficiency of Steam-Generating Equipment" and it will be sufficient to repeat that the average evaporation obtainable from the coal reaching the

Chicago markets is only about 5.5 lbs. of steam.   Higher results
are to be had in some cases, but on account of conditions met
with in the average boiler-room, the general efficiency of production is much below expectations.   There are extensive standby
losses incidental to operating boilers at various and fluctuating
loads during the twenty-fours of the day, and the conditions
are generally not as favorable as are obtainable in electric plants
on the average.   It may be cited that six plants in Chicago
in which the entire boiler output is metered, showed evapora-
tions ranging from 5.0 lbs. to 6.4 lbs. with efficiencies from 44.7%
to 56.7%.   This difference was due primarily to the furnace
design.   For small plants of only a few hundred horse-power,
the evaporation should be assumed at 5 to 5.5 lbs., and for the
larger plants 5.5 to 6.0 lbs., using bituminous coals.

Substituting these values in equation (2) we have

$$T = \frac{S \times 2 \times (70° - 39°) \times (243 \text{ days} \times 24 \text{ hours})}{e \times 1000 \times 2000} = \frac{S \times .1808}{e}.$$

For estimating work it is well to increase this factor by about
ten per cent making, in round numbers.

$$T = \frac{S \times K}{e} = \frac{S \times .2}{e}. \quad \cdots \cdots \quad (3)$$

It must not be supposed that this formula will apply in
every case of heating work, but it may be used with comparative
accuracy in the great majority of estimates and is the simplest
and most easily applied formula yet proposed.   While there
have been published a number of other formulæ and rules,
doubtless correct, yet these will be found to involve so many
variables that the use of them is difficult to any but the most
expert engineers.   In view of the many opportunities for errors
of judgment, on the part of the estimator, the use of this formula,
after once being worked out for any particular locality, will
probably produce results as accurate as more extended calculations necessary in other formulæ.   This formula has been found
to give good results in Chicago and New York City.   The tests
are confined to a comparatively few buildings for the reason

that in most buildings steam heating is only one of a large number of uses for steam, and it is usually difficult to accurately differentiate the direct heating from the total steam consumption, unless it is metered separately. This condition is found where a portion of the exhaust from various units in the plant, such as pumps, engines and compressors, is discharged into the heating system.

It must be borne in mind, however, that the coal consumption as shown by this method could be reduced materially for the reason that it is figured on a flat-rate basis and a liberal margin of safety must of necessity be allowed by the heating company. If steam is to be metered or if the building owner operates his own plant, a saving of fifteen per cent could reasonably be expected.

Method of Application. To illustrate the method of applying this formula, a definite example will be taken,—an office building with the following dimensions:

DATA ON BUILDING, FIGURED FROM PLANS:

	Ft.
Width, east and west..................	202
Depth, north and south................	202
Height, above street...................	260
Basements...........................	28
Number of floors (not including basements)	21
Number of floors (including basements)...	22½

Light court, 73 ft. × 73 ft., above second floor.

	Cu.ft.
Cubical contents......................	9,436,600
Cubical contents (including basements)....	10,293,500

	Sq. ft.
Total gross floor area.................	730,300
Total gross floor area (including basements)	791,510
Equivalent glass area (see below) formula.	125,000
Actual radiation......................	85,000

The above building is exposed on all elevations. The following are details of the estimate on the glass and wall area of the building:

		Sq.ft.
North glass		19,339
West glass		19,206
South glass		6,481
East glass		19,206
Skylight and sidewalks		8,940
Light court		25,428
	Total	98,600
North wall		37,685
West wall		37,728
South wall		50,643
East wall		37,728
Court and roof		54,176
	Total	217,960
Total glass area		98,600
0.1 net wall area		21,796
10% N. and W. glass		3,854
10% of N. and W. wall equivalent		754
	Total equivalent glass area	125,004

In low buildings where the roof forms a considerable portion of the exposure, allowance should be made on the same basis as wall area. If there is an attic space, which is unheated, the roof area should be figured as 0.5 of an equivalent amount of ordinary wall. In high buildings the roof area represents such a small percentage of the total wall and glass area that it may be omitted, unless the attic is heated.

The next step will be to substitute the exposure S, just found, in the formula, $T = \dfrac{S \times K}{e}$, which gives

$T = \dfrac{125004 \times .2}{5} = 5000$ tons of coal per heating season of 243 days which amount will be found sufficient unless there is indirect heating for certain parts of the building. The above amount represents the normal coal consumption and has been verified in this case by observations and records covering several years' operation.

Several formulæ have been worked out for figuring the theoretical radiation required in a building. In Chicago it has been found that the average practice of architects in large office buildings corresponds to about three-fifths of the equivalent exposed glass surface.

$$R = \frac{3S}{5}. \qquad \ldots \ldots \ldots \quad (4)$$

The above formulæ apply only to buildings heated by direct radiation and if there is also to be a system of indirect, or fan-blast heating for the purpose of ventilation, it is necessary to make a separate calculation of the latter requirements.

As has been seen under " Ventilation," the indirect method of heating and ventilating will require more than twice the amount of steam used to operate a direct-radiating system with natural ventilation. Among buildings supplied with this equipment are found: hospitals, schools, libraries and public buildings. The most modern office buildings with two or more basements are required by law to provide certain amounts of ventilation and as the outdoor air must be tempered and brought up to over 70°, the cost of so doing must not be neglected. Department stores and restaurants are required by law to provide their patrons with certain specified amounts of ventilation, and in some high-class restaurants, advantage is not infrequently taken of the fact that a system of indirect heating for winter use may, with slight alterations each season, be converted into a system for cooling the air during the warm months. Accordingly, one of the principal items of information an estimator should obtain is the extent of the hot-blast heating in a building.

The rule for estimating the coal required for this purpose is as follows:

$$T = \frac{Cu. \times (t_1 - t_2) \times H \times D}{55000 \times 2000 \times e} \quad \cdots \quad \cdots \quad (5)$$

where

T = Tons of coal per year.

$Cu.$ = Cu.ft. of air heated per hour.

$t_1 - t_2$ = Difference in temperature between heated air and outdoor air.

H = Number of hours per day during which system is operated.

D = Number of days per season during which system is operated.

e = Evaporation of steam per pound of coal.

Several years ago Mr. Tweedy, whose investigations have already been referred to, made an exhaustive study of the heating requirements of a large number of buildings located in New York City, and as a result he proposed a formula similar to the one used in Chicago. It was found very desirable as previously explained to eliminate the variable factor for air change in all possible cases since this avoids much complication and uncertainty in estimating. Perhaps the formula most generally used to determine heat losses from a building is as follows:

$$\text{B.T.U. per hour} = \left(G + \frac{W}{4} + \frac{nC}{55} \right) \times t, \quad \cdots \quad (6)$$

in which

G = Area of glass in square feet.

W = Area of wall in square feet.

n = Number of air changes per hour.

C = Volume or cubical contents in cubic feet.

t = Difference in temperature between inside and outside.

By considering the last term of this equation as a function of the window area, as was done in the derivation of the Chicago formula, the following expression was adopted:

$$\text{Tons of coal per year} = \left(\frac{G + \dfrac{W}{4.5}}{K} \right) + \frac{G}{K^1}, \quad \cdots \quad (7)$$

where

G = Area of glass surface in square feet.
W = Area of wall surface in square feet.
K = Constant determined for direct heat loss.
K^1 = Constant for air leakage.

By comparing the actual coal consumption of several buildings, values were found for the constants K and K^1, and for the conditions in New York City, it was discovered that no serious error resulted in the application of the formula, if K was assumed to equal K^1. Therefore, in practical application the formula becomes,

$$T = \frac{2G + \dfrac{W}{4\cdot5}}{K}. \qquad \ldots \ldots \quad (8)$$

In comparing this with the Chicago formula it will be noted that the same result is arrived at although in a different manner. In the latter case the exposure is represented by, $S = G + \dfrac{W}{10}$ which may also be written in the form, $S = \frac{1}{2}\left(2G + \dfrac{W}{5}\right)$ from which it is seen that further multiplication by the factor 2, reduces the expression to approximately the same value as used in New York. These expressions in the parenthesis may be considered equal in both formulæ, but the values of K in each case are at variance due to differences in the many factors which have been enumerated earlier in the chapter. In New York, the value of K is as follows:

For office buildings. . . . . $K = 100$
For apartment buildings. . . . . $K = 60$

For Chicago conditions these values in the Tweedy formula would be approximately,

For office buildings. . . . . $K = 50$
For apartment buildings . . . . $K = 35$ to 40

This difference is to be explained by the climatic conditions, temperature and wind velocity and the use of a poorer grade of coal in the latter city. To quote Mr. Tweedy:

" It is evident that this equation cannot be employed with any high degree of accuracy in the case of any building that is so peculiar in its construction as to widely differentiate it from others of its class; also, where mechanical ventilation is to be employed, a certain allowance must be made for the loss of heat arising from this source, over and above that which the equation provides for in the form of natural ventilation due to air leakage. However, as a means of arriving at the amount of coal required to heat a given building, the writer has found this method to be of considerable value, and, owing to the many uncertainties that are invariably present in any heating problem, the use of this method will probably lead to results quite as accurate, in the average case, as those secured by the use of a much more involved and elaborate method of calculation."

Estimating Small and Medium Sized Buildings. District heating companies operating largely in residence districts are enabled, due to the similarity between buildings of this class, to approximate with fair accuracy the probable cost of heating any given residence or store mainly through comparative methods. Experience will serve as the best guide and within reasonable limits very general calculations, such as a determination of the cubical volume and glass surface, will enable the solicitor to compare the results that have been recorded in other buildings of the same size and construction. In the absence of these comparative data, which apply where the heating company is new in the field and has little, if any, opportunity for such comparisons, it will be necessary to make some theoretical estimate of the cost of service. This may be done in the following way, viz.:

1. Determining the heat losses from the building and the cost of an equivalent amount of heating medium, at the proper rates.

2. Determining the square feet of radiation theoretically required and estimating the condensation per season based

upon data known to apply in similar climates and type of construction.

Tables III and IV show in condensed form the variations to be expected in different classes of small and medium-sized buildings supplied on the meter basis, emphasizing among other points that too much dependence cannot be placed on relations between cubical volume and steam consumption. On pages 35 and 36, Chapter II, will be found a typical formula with table of constants for applying the second method above referred to.

As has been noted in previous chapters, the cost of heating service depends so much upon the attitude of the individual customer, and his method of economizing, that it is extremely difficult for an estimator to predict with great accuracy just what the heat consumption will be. Perhaps many of the points brought out in the preceding sections will be of assistance also in estimating small buildings.

TABLE III.

Large City in the Middle West

Mean Temperature during Heating Season 41° F.

No. of Consumers.	Class of Building.	Cubic Feet of Space.	Square Feet of Radiation.	Total Condensation per Season, Lbs.	Season's Average per Thousand Cubic Feet of Space. Lbs	Season's Average per Sq Ft. Radiation. Lbs
20	Office bldg.	6,270,000	77,548	49,503,000	7,895	638
43	Retail store	3,140,000	29,653	11,960,000	3,808	403
50	Residence	2,014,000	40,394	14,882,000	7,390	368
18	Saloons	552,000	6,734	4,175,000	7,563	620
6	Hotels	1,603,000	18,642	10,843,000	6,764	582
2	Apartments	293,000	5,657	3,133,000	10,693	554
4	Garages	798,000	4,284	3,433,000	4,302	801
8	Light mfg.	3,077,000	27,880	9,435,000	3,066	338
9	Wholesale store	2,487,000	14,598	8,534,000	3,431	584
2	Clubs	271,000	4,452	1,298,000	4,789	291
162	Total	20,505,000	229,842	117,196,000	5,715	510

TABLE IV.

STEAM CONSUMPTION OF VARIOUS CLASSES OF BUILDINGS IN A CITY OF THE MIDDLE
WEST, INDICATING THE ECONOMY EFFECTED BY METER RATES

Meter Rate

No. of Consumers.	Class of Building.	Cubic Feet of Space.	Square Feet of Radiation.	Total Condensation per Season, Lbs	Season's Average per Thousand Cubic Feet of Space. Lbs.	Season's Average per Sq. Ft. Radiation. Lbs.
4	Residence	97,950	1,860	1,031,000	10,526	554
12	Business	1,789,066	15,641	8,226,000	4,598	526
1	School	278,085	3,219	605,000	2,176	188
1	Church	194,225	2,202	339,000	1,744	154
1	Hotel	119,750	1,543	1,487,000	12,417	963
5	Bus. and off.	481,776	5,527	4,225,000	8,769	764
3	Offices	681,598	8,624	6,115,000	8,971	709
1	Bank	194,369	2,125	1,560,000	8,026	734
2	Saloons	57,580	1,126	655,000	11,318	581
2	Lodges	807,550	5,706	2,629,000	3,256	460
3	Factories	1,447,935	10,273	6,123,000	4,229	596
3	Theatres	563,344	3,904	1,123,000	1,993	287
1	Miscellaneous	389,510	3,382	3,903,000	10,020	1,153
39	Total	7,103,028	65,132	38,021,000	5,352	584

Flat Rate

9	Residence	209,794	4,039	3,699,000	17,615	917
31	Business	2,720,031	18,608	19,345,000	7,115	1,039
1	School	299,000	2,752	4,200,000	14,020	1,526
1	Church	186,500	593	616,000	3,301	1,039
1	Hotel	79,174	798	1,292,000	16,351	1,621
11	Bus. and off.	1,716,614	16,053	14,779,000	8,610	.921
13	Offices	1,156,289	19,348	14,816,000	12,809	766
9	Bus. and res.	348,685	4,013	3,594,000	10,301	895
3	Banks	139,359	2,302	1,634,000	11,710	710
8	Saloons	243,384	3,233	2,904,000	11,930	899
2	Factories	221,285	960	1,135,000	5,135	1,181
2	Theatres	33,350	382	682,000	20,450	1,785
2	Restaurants	96,682	1,114	1,160,000	11,990	1,040
2	Miscellaneous	409,210	3,867	4,462,000	10,904	1,153
95	Total	7,859,357	78,062	74,318,000	9,456	951

TABLE V.
EASTERN CITY
Mean Temperature during Heating Season 41.3° F.

No. of Con-sum-ers.	Class of Building.	Cubic Feet of Space.	Square Feet of Radia-tion	Total Conden-sation per Season, Lbs.	Season's Average per Thou-sand Cubic Feet of Space. Lbs	Sea-son's Aver-age per Sq. Ft. Radia-tion. Lbs.
122	Residence	2,391,644	40,279	25,217,000	10,543	628
37	Business	1,255,216	11,720	10,450,000	8,335	891
4	Schools	831,117	12,396	4,744,000	5,712	383
3	Churches	473,516	2,220	1,752,000	3,701	790
1	Hotel	291,194	3,505	2,609,000	8,959	745
2	Flats	276,464	3,326	3,265,000	11,809	981
1	Post office	184,580	3,401	1,153,000	6,249	339
1	Y. M. C. A.	109,445	2,200	1,336,000	12,212	607
2	Banks	93,436	969	930,000	9,946	960
6	Offices	595,289	8,570	5,472,000	9,196	638
4	Factories	477,408	2,885	3,549,000	7,439	1,230
2	Lodges	101,914	876	877,000	8,605	1,000
1	Hall	131,350	1,534	1,467,000	11,165	956
1	Printing	46,080	662	428,000	9,304	646
187	Total	7,258,293	94,993	63,249,000	8,714	665

TABLE VI.
LARGE SOUTHERN CITY
Mean Temperature during Heating Season 54.7° F.

No. of Con-sum-ers.	Class of Building.	Cubic Feet of Space.	Square Feet of Radia-tion	Total Conden-sation per Season, Lbs	Season's Average per Thou-sand Cubic Feet of Space. Lbs	Sea-son's Aver-age per Sq. Ft. Radia-tion. Lbs.
11	Residence	573,000	8,590	3,123,000	5,450	363
107	Business	10,634,000	65,880	22,895,000	2,153	347
3	School	632,000	7,414	2,485,000	3,932	335
11	Office bldg.	6,433,000	48,988	25,703,000	3,995	525
8	Hotel	2,461,000	25,444	10,325,000	4,196	406
3	Clubs	745,000	5,922	4,119,000	5,528	696
1	Post office	438,000	5,509	2,541,000	5,801	462
2	Telephone Co.	382,000	3,821	2,169,000	5,678	567
1	Hospital	248,000	4,491	3,147,000	12,690	702
1	City Hall	980,000	9,777	3,471,000	3,540	355
1	Court House	1,020,000	8,866	4,733,000	4,642	534
149	Total	24,546,000	194,702	84,711,000	3,451	435

CHAPTER VII

ESTIMATING MISCELLANEOUS STEAM REQUIREMENTS IN LARGE BUILDINGS

In addition to heating service the district-heating company may be called upon to supply steam for many other purposes. The owner of the modern first-class building is obliged to provide the highest quality of service for his tenants, and this is accomplished only by the installation of a system of auxiliary apparatus of various kinds, many of which are operated by steam. Among these may be included the following:

1. Hot-water heaters, supplying heated water for industrial or manufacturing purposes, or for domestic uses, such as scrubbing and lavatories.

2. Vacuum pumps, used in connection with steam-heating systems for removing air and condensation from the radiation.

3. Ejectors used in a manner similar to vacuum pumps.

4. House pumps used for elevating the domestic water-supply to the roof of the building where it is stored in a tank.

5. Boiler-feed pumps.

6. Steam-hydraulic elevator pumps.

7. Direct-steam elevator engines.

8. Fire-pumps.

9. Air compressors.

    (a) General use.

    (b) Sewer-ejector system.

    (c) For pressure-tanks on hydraulic system.

10. Steam syphons and jets.

11. Refrigerating Machinery.

    (a) Compression systems.

    (b) Absorption systems.

12. Brine pumps or other auxiliary refrigerating apparatus.

13. Drinking water pumps.

14. Stoker engines.

15. Ventilating fan-engines.

16. Warming and cooking apparatus for restaurants.

17. Laundry apparatus.

18. Miscellaneous industrial uses, varying with the class of building and tenants.

To determine with any degree of certainty just how extensive the use of steam for these purposes will be in each building, usually requires a specialist in this particular line of engineering—one who is enabled to draw from experience and observation for verification of estimates. Whenever possible, it is needless to say one should be guided by comparison with buildings already supplied with approximately similar service. In the absence of any such comparative data the following suggestions should be found of assistance.

(1) All modern buildings supply hot water for lavatory and scrubbing purposes; also for barber-shops and doctors' and dentists' offices when required. The most common method is to install a coil heater (similar in principle to a surface condenser) in a convenient part of the basement, the shell or tank enclosing the steam-coil being connected to the main water-supply system and the coil to a steam-supply line in the same manner as direct radiation. Where the building requires no steam for other purposes than heating water, it is the custom to install a small coal or coke boiler for heating the water. Many of the largest office buildings using central-station electric service, but operating their own heating boilers require only a moderate-sized vertical boiler (30 to 50 H.P. rating) to carry the entire load during a considerable portion of the year.

The amount of steam required for heating water varies through such wide limits that it is impossible to give any rule applicable to all conditions. The average of a number of modern office buildings results in a value of approximately 8 lbs. of steam per square foot of gross floor area per year, but the individual buildings varied from less than 5 lbs. to more than

13 lbs. per square foot. A formula which is sometimes used is as follows:

$$\text{Tons coal per year} = \frac{\text{Square feet gross floor area} \times K}{e}. \quad (1)$$

where

K = A coefficient based on experiment ranging from 0.003 to 0.006

e = pounds of water evaporated per pound of coal.

In estimating the requirements for hotels, from 10,000 to 20,000 lbs. of steam per guest room per year should be allowed. For other classes of buildings more exact data must be obtained, if possible the actual quantity of water used being obtained and the amount of steam computed thereon, approximately as follows:

$$\text{Pounds steam per year} = \frac{\text{Gallons water} \times 8.33 \times (T° - t°)}{1000}. \quad (2)$$

in which

$T°$ = temperature of hot water.

$t°$ = temperature of cold water-supply.

(2) When buildings are equipped with the vacuum system of heating, pumps are usually installed to return the condensation to the receiving tank and also relieve the radiation of air as was described in Chapter III. Such pumps are generally driven by steam, but inasmuch as they are in operation only during the period when steam heating is required and as practically all the exhaust steam from them may be introduced again into the system, it is not necessary to consider such units in calculating the requirements of the building.

If the pump is used for some purposes other than heating, such as returning the condensation from apparatus where no use can be made of the exhaust steam. the only method of calculation is by figuring the displacement of the steam-cylinder and estimating the number of strokes, from which it is possible to obtain the volume and weight of steam used. To the theo-

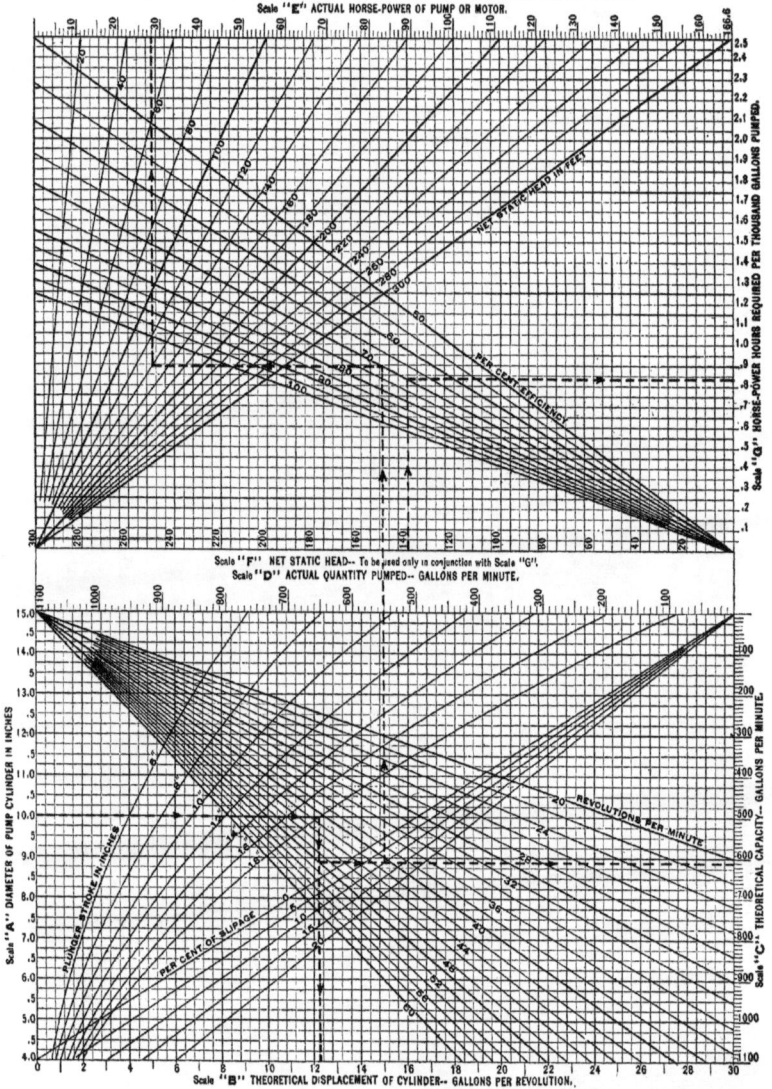

FIG. 69.—Graphical method for solution of pumping problems   *To face p.* 233

retical volume it is advisable to add at least 50% to allow for leakage and other losses in all classes of pumps.

(3) Ejectors, which are used almost exclusively in connection with the " Paul " system of heating may be met with occasionally, and may be assumed to require from 2% to 4% of the steam used in the heating system. It is necessary to supply high-pressure steam to the ejector and this represents a positive loss because the exhaust cannot be utilized.

(4) House pumps usually receive their supply of water through surge-tanks, direct from the city mains and their function is to elevate same to the roof-tank where it is stored and used in a number of ways. Data must be obtained as to the size of the tank and number of times it is filled in any specified period; or, if a water meter is available the total quantity pumped will be easily determined. The next step is to compute the horse-power required. As this involves considerable calculation, a diagram, Fig. 69, is given by the use of which it is possible to solve any problem which may arise, in a very simple manner. The horse-power required may be computed by the following formula.

$$\text{H.P.} = \frac{2 \times 3.1416 \times D^2 L}{4 \times 231 \text{ (cu.in. per gal.)}} \times R \times \frac{100-S}{100} \times \frac{H \times 8\frac{1}{3}}{E \times 33000}. \quad (3)$$

The horse-power hours required, to pump 1000 gallons of water, assuming the head in feet to be known, may be determined as follows:

$$\text{H.P. hours} = \frac{H \times 8\frac{1}{3} \times 1000}{E \times 33000 \times 60}, \quad \ldots \quad (4)$$

in which formulæ,

D = Diameter of plunger in inches.

L = Length of stroke in inches.

R = Revolutions per minute.

S = Per cent of pump slippage = (5% to 40%) depending on condition of pump.

H = Net static head.

E = Per cent pump efficiency = (50 to 80).

To simplify the use of the diagram the following problem has been worked out thereon as indicated by the broken lines,

Assume in (3) and (4),

$$D = \text{10 inches;}$$
$$L = \text{18 inches;}$$
$$R = \text{50 per minute;}$$
$$H = \text{140 feet;}$$
$$S = \text{10\%;}$$
$$E = \text{70\%.}$$

Substituting in (3),

$$\frac{2 \times 3 \cdot 1416 \times 100 \times 18}{4 \times 231} \times 50 \times \frac{100 - 10}{100} \times \frac{140 \times 8\frac{1}{3}}{.70 \times 33000} = 27.8 \text{ H.P.}$$

as shown on Scale E.

Substituting in (4),

$$\frac{140 \times 8\frac{1}{3} \times 1000}{0.70 \times 33000 \times 60} = 0.84 \text{ H.P. hours per 1000 gal.}$$

as shown on Scale G.

After having determined the horse-power requirements the amount of steam must be estimated. The ordinary duplex house-pump requires from 100 to 200 lbs. of steam or more per horse-power hour, probably averaging about 140 lbs. Compound steam-cylinder pumps are about 20% more economical in the use of steam. The efficiency of a pumping unit depends so much upon the manner in which it is maintained that no rule will apply in every case.

The electrically-driven triplex or centrifugal pumps are becoming very popular and nearly every modern building using central-station service is equipped with them.

(5) Boiler-feed pumps may be asumed to require from 1 to 2% of the total output of steam. As this amounts to a small item, and since the exhaust from such pumps may usually be more profitably utilized in heating the feed-water it is not customary to credit them with supplying steam for the heating

system. If desired, the steam consumption may be checked by calculation in the same manner as was described for House Pumps, (4) the steam consumption being at the same relative rates per horse-power.

In some few instances turbine-driven centrifugal pumps have been used and the efficiency of such units is greater than that of recriprocating pumps. Electric boiler-feed pumps have been advocated for some years, but are very seldom used.

(6) Hydraulic-elevator installations are losing popularity in favor of the more modern and less expensive electric equipment now on the market, but in many of the buildings erected up to within a few years ago they were much used. Therefore it will be well to consider the cost of pumping the water for their operation.

Hydraulic elevators are usually classified and designated as per Table I. This table shows also the relative efficiency of each type and the water horse-power hours consumed per car-mile of travel.

TABLE I *

SPECIFICATIONS FOR HYDRAULIC ELEVATORS

Type	Usual Water Pressure. Lbs. per Sq. In	Final Net Efficiency. Per Cent	Water H-P. Hours per Car-mile
(a) Vertical high pressure	580	o 547	6 10
(b) Vertical low pressure	145	o 522	6 39
(c) Horizontal low pressure	150	o 443	7 52
(d) Plunger low pressure	226	o 333	10 00

* Thomas E. Brown—Transactions—A S C.E. 1904—paper 22.

The mechanical efficiency of an elevator is the ratio of the live load lifted to the power expended, whereas the hydraulic efficiency, which is largely a matter of proportioning the piping, cylinders, etc., depends upon the designer. The load efficiency considers the unbalanced weights and varies with the type of elevator. The product of these efficiencies gives the final net efficiency as shown.

It has become the accepted custom to use the term " car-mile " in describing the performance of elevators, steam, hydraulic or electric. In the case of hydraulic elevators the water horse-power hours represent the net work performed upon the water by the pumping-engines and accordingly the final economy of a given elevator installation revolves itself largely into a question of the relative steam rates of the particular pumps used.

In the vertical high and low-pressure type (a and b) the working cylinder (usually about 30 ft. in length) containing the piston rod which gives movement to the shieve-blocks is installed in the elevator shaft in a vertical position near the car. The length of stroke depends upon the gear-ratio which is usually designed for about 6:1 in buildings of approximately 200 ft. in height. The usual pressure carried is 580 lbs. per square inch on high-pressure service although numerous installations, known as accumulator systems, are operated at pressures in excess of 1000 lbs. per square inch. As this high pressure involves heavy expense for maintenance and is liable to accident such systems are not being installed as much as formerly, the gain in efficiency being so slight as to be offset by the disadvantages.

The vertical low-pressure elevator (b) requires only 150 lbs. hydraulic presssure and is the most widely-used type of machine, because of its high efficiency and economy in space occupied by the cylinder.

The horizontal type (c) is but very little used although many of the earlier buildings are so equipped. The principal disadvantage is in the great amount of basement space required. In order to overcome this, the gear-ratio is increased to 12 : 1 which permits a shorter cylinder.

Plunger elevators have been used since about 1900, but are gradually being discontinued. This type is most economical of space, the cylinder being sunk in the ground directly under the center of the car and this form of construction found favor with the public since it was generally believed much safer; the idea prevailing that the car was supported (rather than suspended) from below and that there was no danger from breaking cables. However, considering the high-running speeds now demanded, the plunger type is by no means the most secure, and although the mechanical efficiency is high the large unbalanced weight required to bring the car to a quick stop reduces the final efficiency.

There are two general methods in use for estimating the steam consumption of elevators:

" 1. By determining the amount of water to be pumped, figuring same from the actual or theoretical cylinder-displacement and estimating the water horse-power hours required per trip, per car-mile or any other convenient unit. If the guaranteed steam-rate per horse-power of the pumping unit is known the result is readily obtained, or

" 2. By determining the car-mile travel per year and assuming a certain guaranteed steam consumption per car-mile; or by comparison with actual tests of similar pumps in previous installations."

In large buildings the practice has been to install pumps of the compound Corliss-fly-wheel type, sometimes known as " Grasshopper " pumps. These pumps usually are fitted with one high-pressure cylinder exhausting through a receiver into two low-pressure cylinders. The cost of maintenance is rather high but the economy is far in advance of any other type of pumping machine for a similar purpose. It is claimed they produce an indicated horse-power on 25 lbs. of steam which is equivalent for a pump of this efficiency to 30 lbs. per hourly water horse-power. While these results have been obtained in a number of instances and are guaranteed by manufacturers yet it is advisable to estimate the consumption at not less than 40 lbs. steam per water horse-power, since the continual speed variation following the fluctuating load results in excessive losses from condensation.

The triple expansion, direct-acting or straight-line pumps require 50 to 60 lbs. of steam per horse-power hour; a compound pump, 70 to 90 lbs. and single-expansion duplex-pump, 90 to 120 lbs. per horse-power hour.

The following table is based on the efficiencies given in Table

I and while these values may hold true for new pumping equipment, for ordinary installations an increase of not less than 25% should be allowed, and in case the general conditions indicate very inferior economy perhaps 50% should be added to the values given. This table (with some additions) is computed after extensive tests by the author of Table I.

TABLE II.

POUNDS OF STEAM PER CAR-MILE

Kind of Pump.	Vertical high Pressure Lbs. per Sq. In.	Vertical low Pressure Lbs. per Sq. In.	Horizontal low Pressure Lbs. per Sq. In.	Plunger low Pressure. Lbs. per Sq. In.
Compound fly-wheel pumping engine.........	183	192	226	300
Triple-expansion duplex pump..............	244	256	201	400
Compound, duplex pump...................	427	447	526	700
Simple, duplex pump......................	550	600	700	900

The average travel per elevator in modern office buildings is about 15 car-miles per day. The mileage of elevators in department stores is much lower, averaging perhaps not more than 8 car-miles per day, due to the fact that stops are made at practically every floor. The steam consumption will not be materially lower than for office buildings, as the economy depends upon the number of times the machine is started and stopped. Hotels should be figured on the same basis as department stores, taking all the elevators on the average. Freight elevators are not used to the same extent as the passenger cars, and therefore require somewhat less steam.

In estimating car-mile travel for hydraulic elevators, it must be remembered that no power is consumed on the down trip, hence the power per car-mile is represented by the piston-displacement corresponding to a half-mile run of the car. For a rough and ready rule, where no data of any kind are available it may be assumed that each passenger elevator will consume 2,000,000 lbs. of steam per year. In large buildings where the equipment is more economical the mileage will be greater than in small buildings, but the smaller and consequently less efficient pumps in the small buildings will bring the consumption up to the same basis.

(7) In some of the older buildings, elevators of the direct steam-engine driven-type will be found. These units operate much like an ordinary two-cylinder (both high-pressure) hoisting-engine connecting by means of a worm-gear to the cable drum. It is quite difficult to maintain such apparatus in economical condition, due to excessive valve and piston-leakage, permitting steam to " blow " through into the exhaust. The best method of estimating the steam consumption is to consider the engine as a pump, taking steam full stroke, counting the actual strokes, and multiplying same into the displacement of the cylinder. Liberal allowance should be made in all cases for poor condition of pumps.

(8) Fire-pumps present a very uncertain problem, as it is generally impossible to determine the number of hours they will be in service. These units are usually for emergency service in connection with sprinkler service and operate only when necessary to maintain a certain specified pressure on the system, or in case of an actual fire. Obviously no rule can be given covering the steam consumption, but it should be remembered that it is necessary to maintain pressure up to the throttle-valve of the pump and this entails a certain constant loss from condensation. Present-day practice seems to favor the use of motor-driven centrifugal fire-pumps, such apparatus being easily adapted to automatic-control devices and being the ideal system in many respects.

(9) (a) In nearly every building many uses may be found for compressed air. Among these are the following:

For blowing dust and foreign matter from motors and other apparatus.

For operating the thermostatic control-valves on the heating system.

For the use of doctors and dentists.

(b) Perhaps the most important function of the air-compressor is in connection with the sewage or drainage system. Where air is used for this purpose it is generally for the so-called

" Shone " or air-ejector system by which the drainage products are forced from the sump-pit below the lowest basement level into the street sewer. In large cities, where the sewer facilities are frequently inadequate at times for the enormous service imposed upon them, and in the absence of ejectors, many buildings are flooded at certain periods of the year during abnormal rain fall. The electric bilge-pump is also largely used for this class of service, in which case an auxiliary or emergency steam-pump is sometimes installed in case of any possible failure of the electric pump.

(c) In case hydraulic elevators are used, compressed air is admitted into the pressure tanks where it acts as a cushion, and a similar use is made of it for the sprinkler tanks on fire systems.

Where steam-driven compressors are installed those most commonly found are of the well-known Westinghouse Locomotive type, the steam consumption being determined in the same manner as pumps, i.e., displacement of steam-cylinder.

(10) Only in rare cases will steam-syphons be found in large buildings, and only in case the drainage ejectors are omitted. Buildings having only one basement, higher than the street sewer sometimes dispense with ejectors and rely upon syphons exclusively in case of floods. If it is necessary to estimate for such apparatus, the following well-known formula will apply:

$$W = \frac{60 \times P \times a}{70}, \quad . \quad . \quad . \quad . \quad . \quad . \quad (5)$$

where

> W = Weight of steam per minute.
> P = Absolute pressure of steam, pounds per square inch.
> a = Area of steam orifice, square inches.

(11) Large office buildings, department stores, hotels and many other buildings are equipped with apparatus for producing artificial refrigeration. This is required for a variety of purposes—cooling drinking water, cooling air used in the warm

months for ventilation, ice boxes, in bars, kitchens, etc., and for manufacturing plate or can ice.

Two general systems are now in use, namely

(a) Compression systems.
(b) Absorption systems.

The compression system may be operated either by a steam-engine or electric motor. In the former case the steam consumption may be estimated in the same manner as for a steam-engine and will range from 20 to 40 lbs. per indicated horse-power per hour, depending upon the type and capacity, Refrigerating machines are rated in "tons of refrigerating capacity per 24 hours," or tons of ice per 24 hours. The indicated horse-power per ton of refrigeration ranges from 1.5 in the larger plants to 2.5 in the smaller sizes. In rating these machines the refrigeration rating in tons is twice the actual capacity for manufacturing ice.

Absorption machines may be operated with either live or exhaust steam. This system requires from 30 to 50 lbs. of steam per ton of refrigeration capacity.

(12) Pumps for circulating the brine, cooled by the compressor or generator, to different parts of the building are generally steam driven, and the steam consumption may be figured by counting the strokes or estimating the quantity of the brine pumped. The steam-rate of such pumps is high, 150 to 250 lbs. of steam per horse-power.

(13) Drinking-water pumps are used for circulating filtered water through an independent system of water mains.

(14) Stoker-engines require from 50 to 60 lbs. of steam per horse-power hour. In the absence of any definite data, estimate 1% of the total steam output.

(15) Steam-engines are frequently used for driving fans in indirect-heating systems. The exhaust is generally utilized in the heating coils and the net consumption does not amount to more than 15 or 20%, which represents the loss through the engine-cylinder. If the exhaust is not utilized the consumption

may be estimated at from 40 to 50 lbs. of steam per horse-power hour.

(16) The supplying of steam for restaurants and lunch rooms is very often an important source of income for the district-heating company. Table III gives the approximate rate of steam consumption for various pieces of apparatus, commonly used:

TABLE III.

Kitchen

	Pounds Steam per Hour.
10 gal. coffee urns	30
5 gal. coffee urns	15
Steam table	1.5*
Bain Marie	3.0*
Stock Kettles	30
Vegetable steamers	60
Oyster pots	15
Egg boilers	15
Warming ovens	30
Plate warmers	30
Dish washers (single)	30
Hot-water tanks	30
Potato boilers	60
Soup kettles	30
5 gal. jacketed kettle	15
Dish-washing sinks	30
Small bean warmer	15
Chocolate urn	15
Silver washing jets	90

See also, N. E. L. A. report, 1913, pp. 483 to 486.

(17) Large department stores and hotels usually operate laundries and Table IV indicates the approximate use of steam for such purposes.

* Pound steam per square foot.

TABLE IV

	Pounds Steam per Hour.
Washer, 3 ft. diam................	60
Washer, 6 ft. diam................	120
50 gal. starch kettle..............	60
Dry room per draw................	15
48 in. ×8 ft. o in. mangle.........	60
50 gal. soap kettle................	60
48 in. ×8 ft. o in. double mangle....	120
16 in. cyl. 8 ft. o in. long.........	30

(18) In addition to the above requirements the following may be mentioned:

Steam for dry-kilns in carpenter shops.

Steam for thawing out down-spouts and roof water-tanks in severe winter weather.

Steam driven pneumatic-blowers.

Steam jets for boiler-furnaces, and

Humidifiers.

CHAPTER VIII

RELATION BETWEEN HEAT LOAD AND ELECTRIC LOAD IN BUILDINGS

PERHAPS there is no question connected with the production of light, heat and power, on which there has been expressed more difference of opinion than on the subject of this chapter.

For the past twenty-five years and, in fact, ever since the installation of electric lighting plants in private buildings, the extent of the economy gained by using the exhaust steam from engines for heating the building has been an interesting subject of discussion. Some have claimed that the steam required for heating buildings could be passed through a dynamo and utilized for light and power, thereby making the heat cost a nominal one. Others ignore the fact that low-pressure steam requires nearly as much coal for its production as is required by high-pressure steam and claim that the saving is very small, basing their argument on the fact that it would require from 80 to 100 lbs. pressure to operate the dynamo whereas it would only require about 2 lbs. pressure to heat the building. Many debates have been held on this subject, each side basing their arguments on various instances which they thought tended to establish their case. Inasmuch as in most private plants the steam from the boiler flows not only into the high-pressure pipe which delivers the steam to the engines but also through a reducing-valve into a low-pressure pipe which supplies steam for the heating system, and inasmuch as for a long time there were no accurate data available, the question was regarded by many as one of those subjects which, like politics and religion, are often decided more by the natural attitude of the individual than on the basis of demonstrated facts.

244

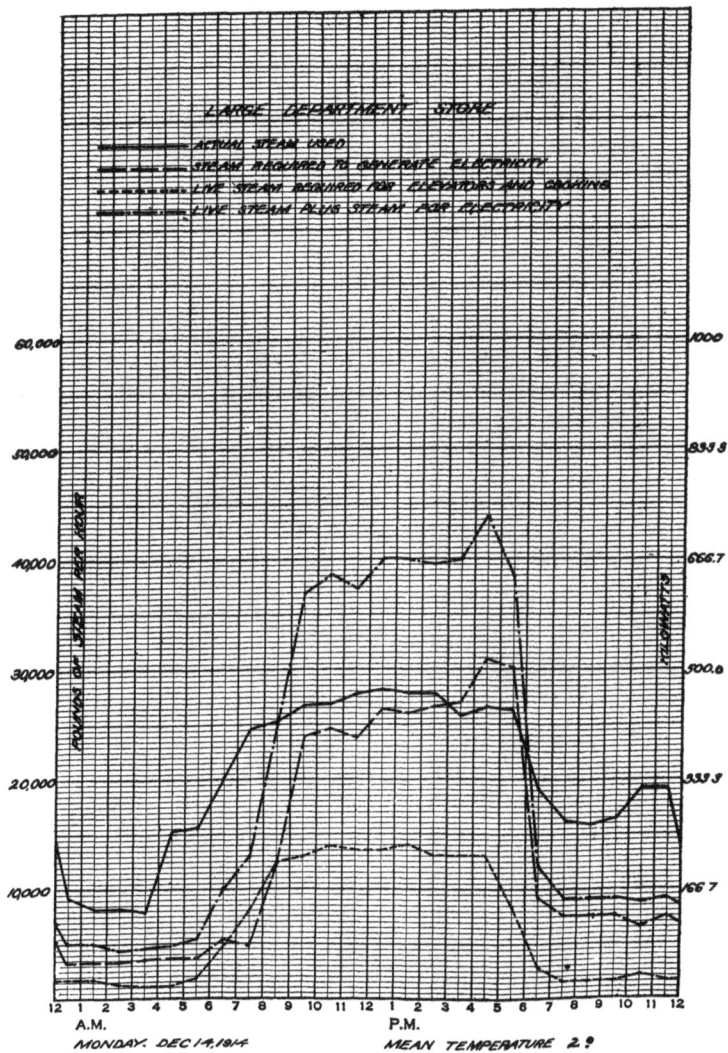

FIG. 70.—Steam and electric load curves, large department store

While it has been known for a long time that outside temperatures in the morning are usually colder than in the afternoon, and on the other hand, that the power and lighting requirements are likely to be heaviest in the latter part of the afternoon, it has been impossible to show how far this lack of correlation of heat and electrical demands would affect the result. One of the first attempts to give a rational solution of this problem was included in a paper given by Mr. Davis Boyden, early in the year 1912, before the American Society of Heating & Ventilating Engineers, in which he took up in a general way the relation between the steam requirements and the electrical requirements of a large department store in Boston. Nearly two years before this, however, some interesting curves were submitted by Mr. A. D. Spencer in connection with central-station systems in Detroit. These curves showed the operation of one of their plants which supplies electricity to the street mains and also supplies exhaust steam to their heating system. They showed that during a large part of the time, electricity had to be supplied to an alternating-current transmission-line in order to make the electrical load correspond with the lighting load. In other words, the curves show a large peak in the morning hours caused by the steam requirements, and another peak in the afternoon caused by the electrical requirements. Quoting from Mr. Spencer's paper, we have as follows:

" From the curves of Figs. 12 and 13, it will be seen that if it were not possible for the heating company to dispose of its electricity outside the district or have load other than lighting load, the plant would become practically a live-steam plant. The plant is now operated as far as possible to supply just enough exhaust steam for heating. As the peak of the heating load comes in the morning when there is practically no demand for light, and as the peak of the lighting load occurs in the afternoon, when as a rule the demand for heating is much less, it is necessary to convert a large part of the direct-current generated to alternating-current and transmit it elsewhere for use, also at the time of the lighting peak it generally is necessary to import and convert a considerable part of the current unless it is desired

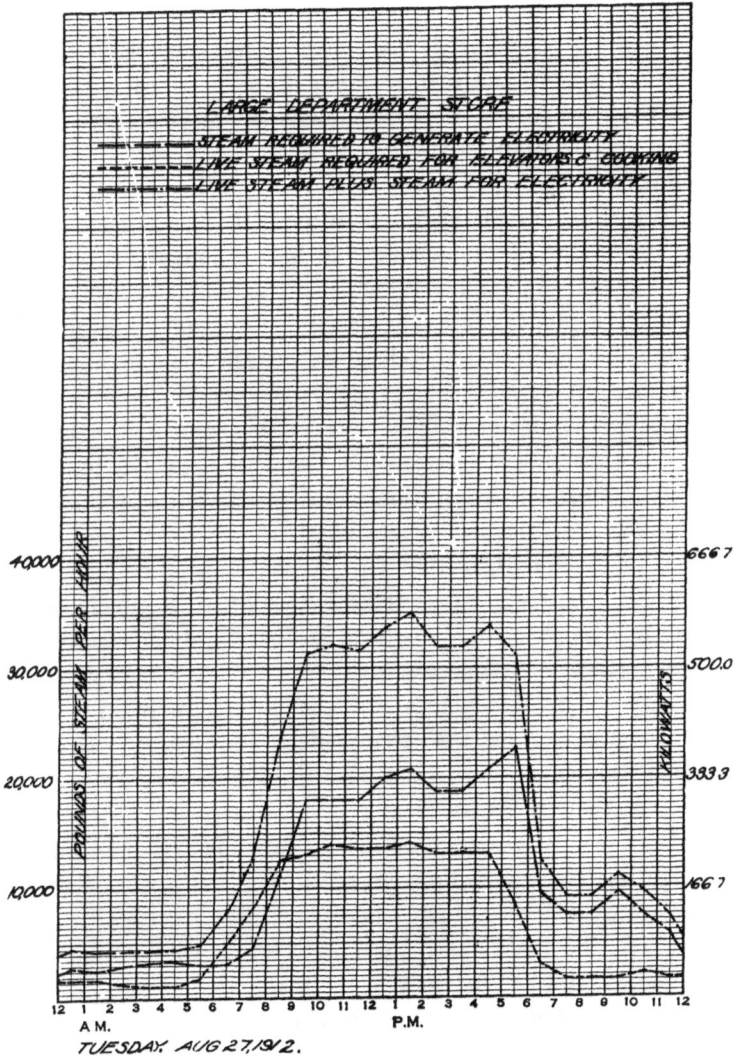

FIG. 71.—Steam and electric load curves, large department store

to run the plant and exhaust steam into the atmosphere which
would be more costly than to bring in the current from the
outside.

" Again a considerable part of the electricity generated is
lost in conversion and in transmission and at the time of electrical peak when the capacity might be most valuable, the
plant is seldom running at capacity."

In conclusion Mr. Spencer states:

" Where a plant is operated in parallel with a much larger
plant, the value of the current will be low and the advantage
will be small. When the plant is not operated in connection
with a large plant, the available electricity will be insufficient
to supply exhaust steam, unless the heating load be comparatively small."

It is further interesting to note in this same report on the
cost of using steam in Detroit, that the actual cost per thousand
pounds of steam in the station which produces steam only and
did not have any electric generating-plant was only a very little
greater than the cost of producing steam for heat in the station
where the steam was also used for providing electrical power.
An attempt has been made both in St. Louis and Chicago by the
central-station companies in a very small way to generate
electricity by isolated plants and feed it into the main central-
station system and use the exhaust steam for heating. In order
to make this experiment successful, it is necessary to operate
the plants just up to the heat requirements. In other words,
the plants feed just enough electricity into the systems to make
the lighting load simultaneous with the heat load. This is
possible only in cases where there is a large electrical system
into which the plants may feed and, while the arrangement
helps the showing of efficiency in the plant which furnishes both
steam and electricity, the arrangement militates to a certain
extent against efficiency in the central-power stations.

In the year 1912 this subject was taken up by the steam
committee of the National Electric Light Association. A series
of curves were obtained by members of the committee from a
number of typical plants. These curves show the actual steam

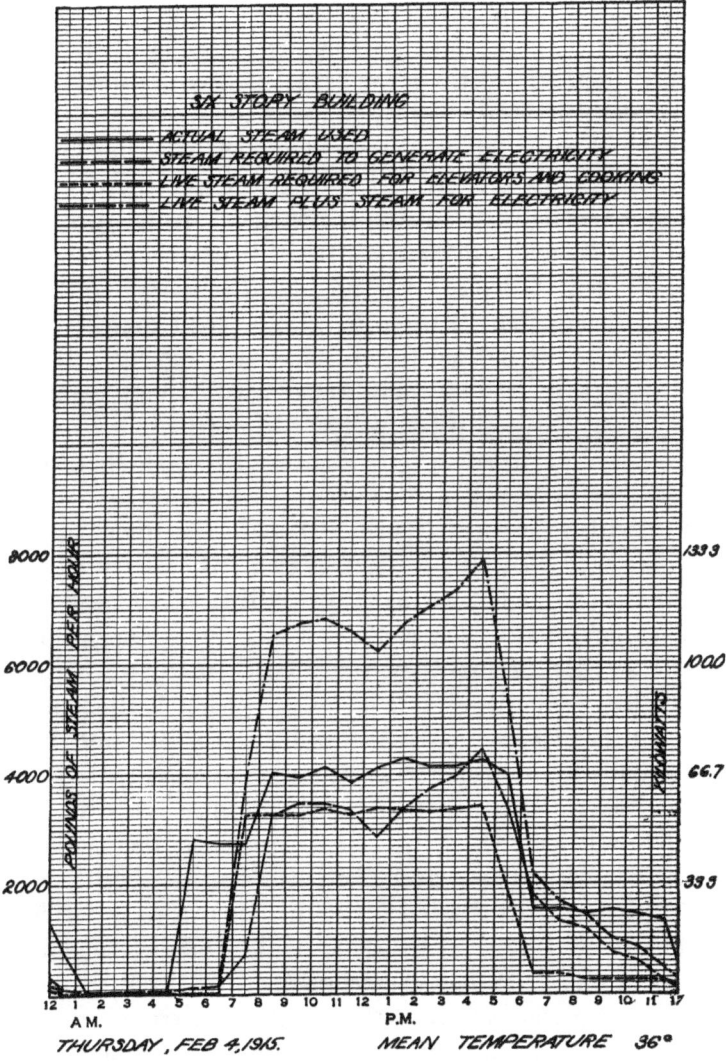

FIG. 72.—Steam and electric load curves, six-story building

and electrical requirements of those plants on various days during the year. These curves were obtained by measuring the amount of steam used in certain plants which were using central-station service and plotting the steam used in these buildings from the records of the St. John curve-drawing meter. On this same drawing was plotted the simultaneous load curve showing the electrical load required in the same building at different hours of the day. In order to reduce the electrical curve to its equivalent in steam requirements, it was necessary to assume the number of pounds of steam that would be required per K.W. hour, provided an isolated plant were furnishing this service. The matter was taken up with different engineers and at first it was thought that the ordinary single-expansion engine direct connected to a dynamo would make a unit requiring about 50 lbs. of steam per K.W. hour. In order to make sure, however, this was checked by a number of tests in New York, Chicago, and other places in plants of actual operation. These tests showed a considerably higher average, viz., about 60 lbs. per K.W. hour. In view of the fact that most plants need to operate at night under very light loads, which would mean very low efficiency, and in view of the fact that the average plant is not new, but has been operating a number of years, it was decided that 60 lbs. would be a fair assumption. While it is doubtless true that the efficiency of a small plant would vary from light load to full load, it was not thought best to complicate the problem by trying to compensate for this difference. The average efficiency was assumed as the same throughout the 24 hours.

These investigations have gone a long way towards clearing up the matter under discussion but there is still room for considerable investigation along these lines. Different buildings vary greatly in their load characteristics. The ordinary high-class theatre giving one, or at the most, two entertainments per day, would have an entirely different condition from the modern hotel which maintains a very much steadier demand for both light and power. These curves showed that while there was a considerable amount of economy in the use of exhaust steam for heating, the extent of this economy had often been exagge-

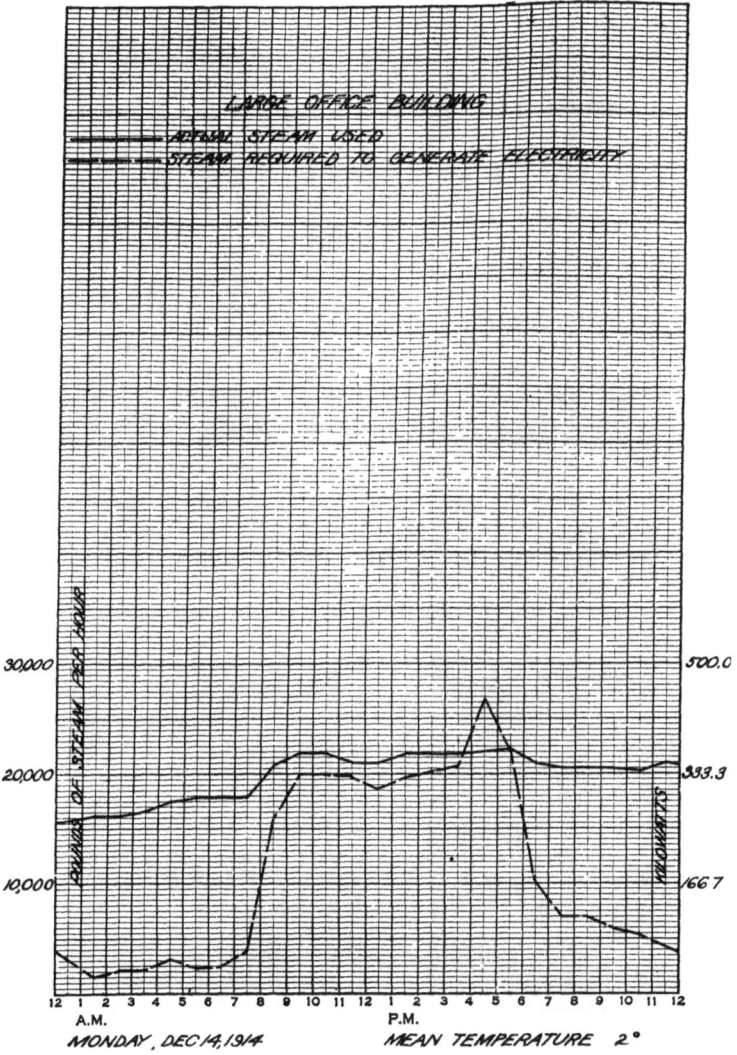

FIG. 73 —Steam and electric load curves, large office building

rated.   It should be remembered in this connection that these curves show the condition only during the cold weather.   During the summer time there, of course, can be no saving as the premises do not require to be heated and during the spring and fall the amount of heat required is comparatively small.   In order to show the nature of these comparisons there are given herewith a number of curves illustrating the relation between the heating and electrical load.

Fig. 70. Shows the simultaneous steam and electric curve for a large department store in Chicago, on a severe winter day.   This store requires a considerable amount of live steam in the summer time for operating a large bank of elevators and for cooking purposes.   The steam curve has been determined by various meter readings taken in the summer and the electric curve is therefore super-imposed on this steam load.   While this introduces a certain amount of estimating in computing the curve, the requirements of the store are pretty well understood and the comparison between the steam and electric curves may be relied upon as practically correct.

Fig. 71. Shows the steam load of this particular store on a summer day, also the electrical requirements.   The curve showing the steam requirements is used as the basis for figuring the electric curve for the winter load.   The space between the steam curve and highest curve will show the amount of exhaust steam which would be necessarily wasted during the summer months if a private plant were in operation.

Fig 72. Shows the conditions of a six-story building in Chicago, which also uses steam for elevator service but purchases electricity for lighting and general power.   In this figure, the electric curve is super-imposed on the summer steam requirements in the same manner as in Fig. 70.

Fig. 73. Shows the steam and electric curve during a very cold winter day in a large office building, which is exposed on all sides.   This curve shows that in exceptionally cold weather the steam requirements for heating reach very nearly to the steam requirements for electricity throughout the day.

Fig. 74. Shows the conditions in the same building on a typical day during moderate winter weather.   This curve shows quite a change in the relation between the electric and steam loads and is also on a day which might be considered an average during the winter season.   The summer load curve of this building is not given as there is practically no steam used in the summer time.

Figs. 75 and 76. Show the conditions in a moderate sized office building located on the corner and practically exposed on all sides.   Fig. 75 shows the conditions in extreme cold weather, while Fig. 76 shows a typical winter day.   In this building also, the steam required n the summer time is practically negligible.

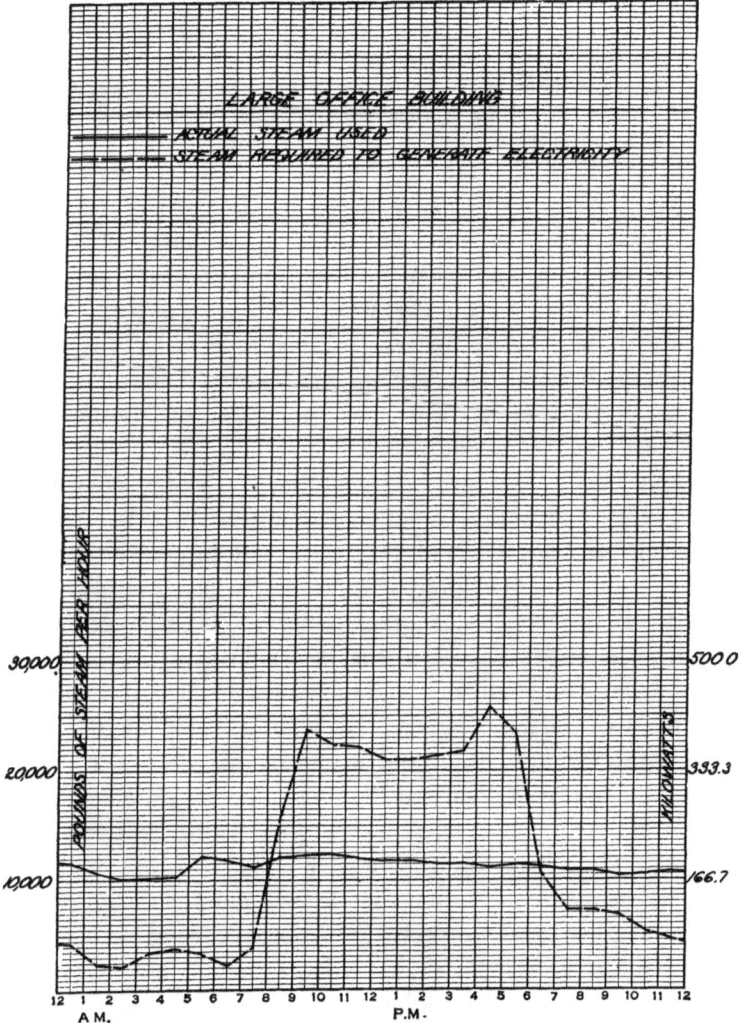

FIG. 74.—Steam and electric load curves, large office building

Fig. 77. Shows the conditions in a large city club which uses electricity for lighting and power throughout the building, with steam for heating, cooking and ice machinery. This curve represents the conditions on a typical winter day. In order to get the requirements, the electric curve is super-imposed on the summer steam load curve which shows the summer live steam requirements.

Fig. 78. Shows steam and electric curves in a modern hotel recently erected, in which electricity is used for light and power, with the exception of a few small pumps and ice machinery.

These curves have been made up by the method described above, as used by the heat committee of the National Electric Light Association, and most of them are plotted from printometers and compared with the steam load shown by a curve drawing St. John steam-meter. In Fig. 78, however, the steam readings are based on the readings of a Venturi-meter. In Figs. 71, 75 and 76, the electrical curves are based on hourly watt-meter readings throughout the 24 hours.

These curves represent the simultaneous values on typical winter days. In order to get an exact relation between heat and electrical load, it would be necessary to integrate the total electrical and steam requirements for each day and also integrate the simultaneous steam requirements. If this were carried through the entire winter season, it would give us the exact relation for any particular building. While such information requires a considerable amount of labor it would undoubtedly be interesting information and lend valuable assistance in the problem of analyzing Isolated Plant conditions. A general idea of the result can, however, be obtained by taking the average of several typical curves during winter conditions and assuming same as the average condition in the five months during which winter weather can be expected; viz., from November to March inclusive. It is recognized that a large amount of information and data must be had in order to give a comprehensive view of this somewhat intricate problem. These curves are submitted, therefore, not as conclusive but simply as additional evidence on this subject. It has been found by experience in Chicago, that the annual bills for steam heating in large buildings are from one-fourth to one-third the annual cost of operating and main-

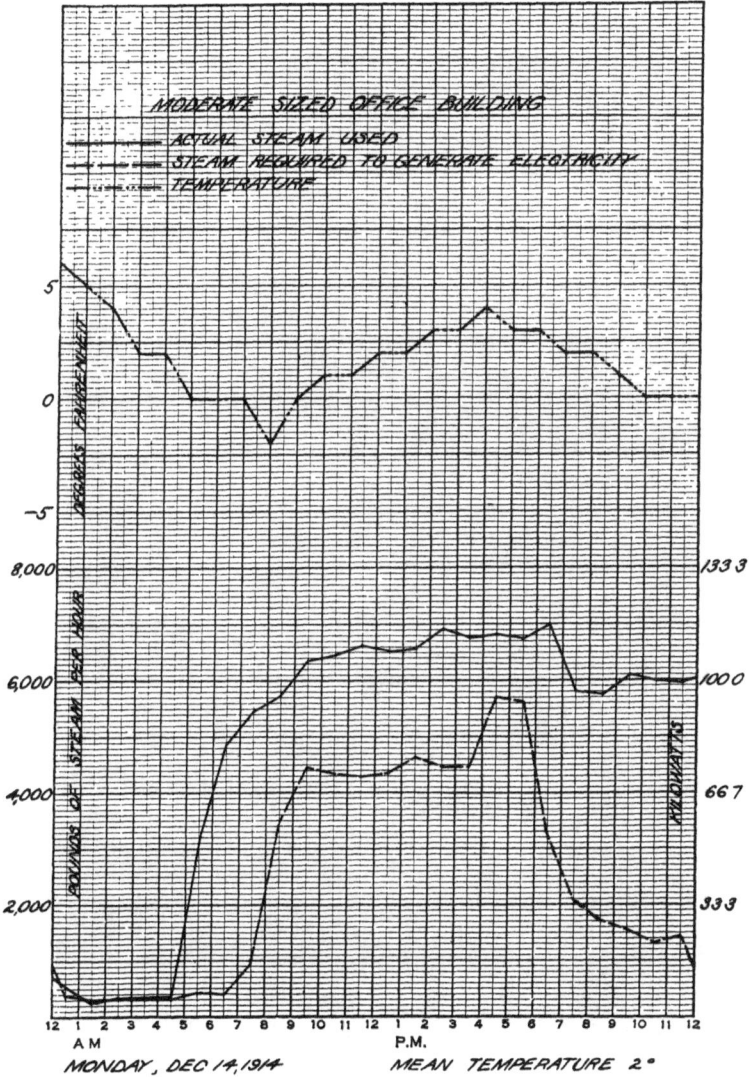

FIG. 75.—Steam and electric load curves, moderate-sized office building

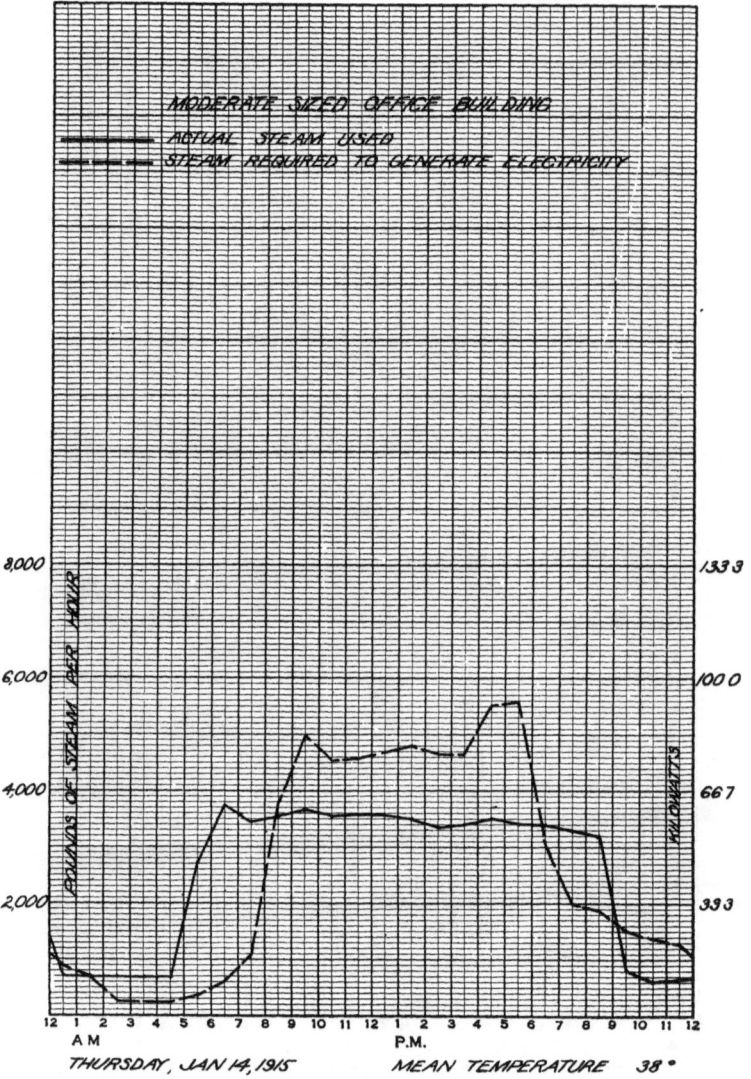

FIG. 76.—Steam and electric load curves, moderate-sized office building

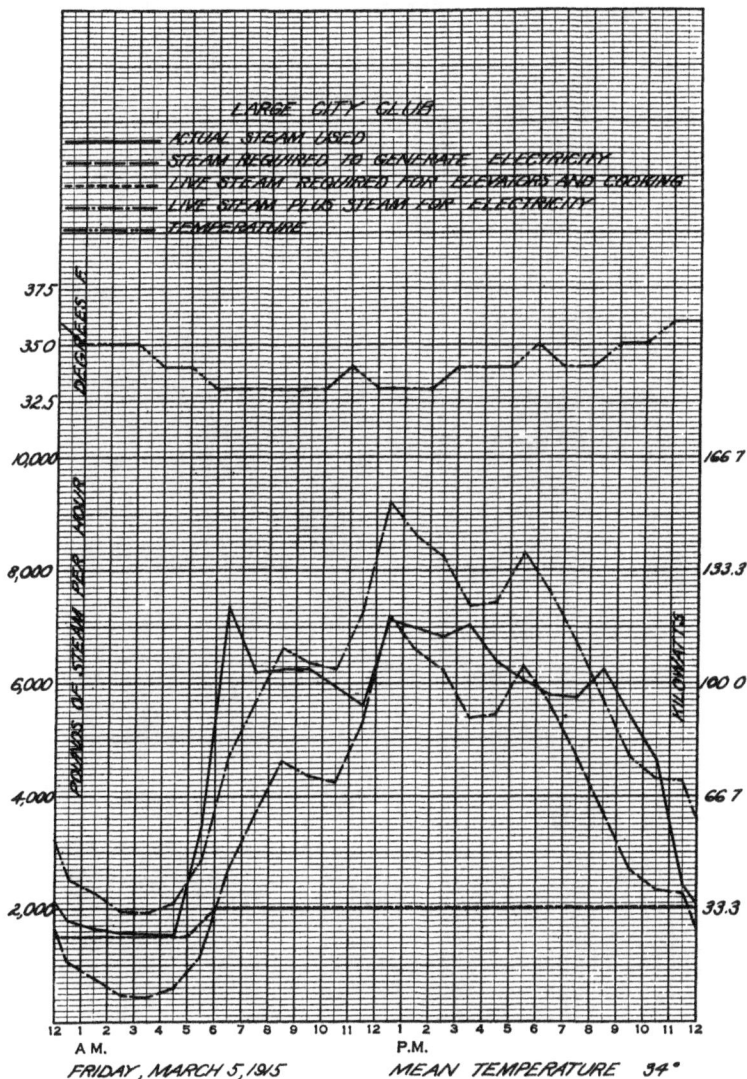

FIG. 77.—Steam and electric load curves, large city club

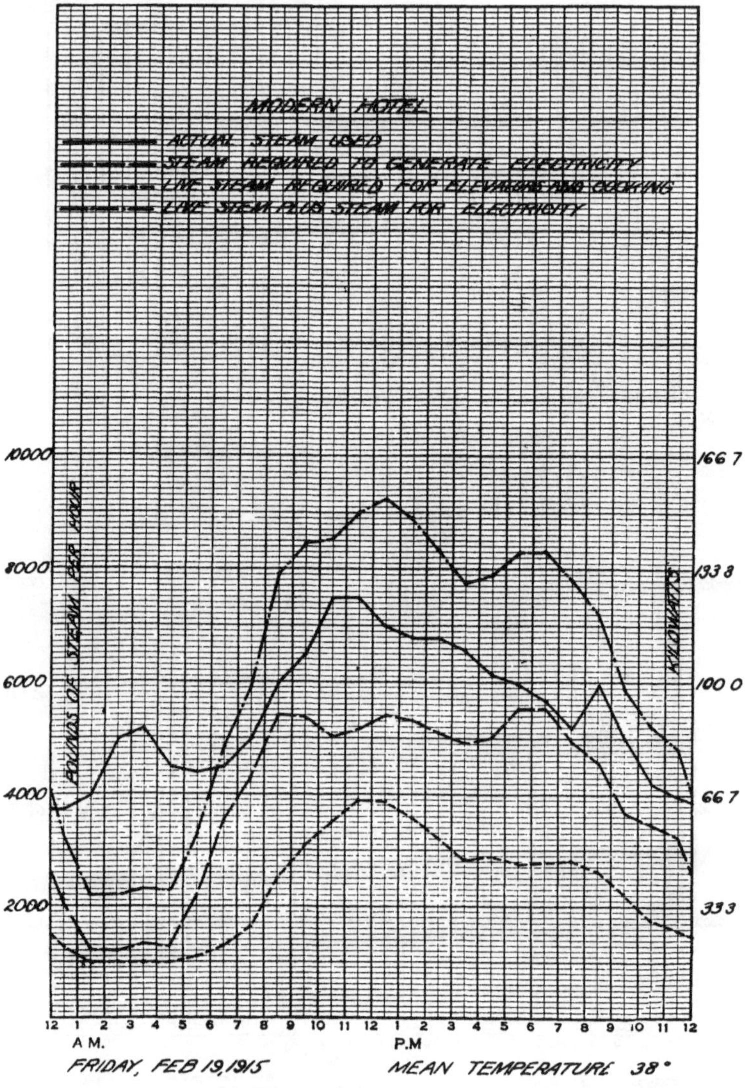

FIG. 78.—Steam and electric load curves, modern hotel

taining an independent isolated plant. Accordingly a large percentage of saving in the cost of steam heating would be a relatively small factor in the total cost of maintaining and operating a plant. For example, if 60% of the steam required for heating a building could be utilized in operating dynamos, the saving as compared with the total cost of heat and electricity would be not over 20%.

The curves given in this chapter are taken from buildings located in the city of Chicago, and in many of the cities of this country the extreme low temperatures herein shown would not be experienced. It is evident that while this is a subject on which it is unsafe to make dogmatic assertions, enough has been shown to prove that the central-station man, on the one hand, should recognize that the use of exhaust steam for heating constitutes an important economy in the operation of isolated plants, but on the other hand, the plant operator should recognize that there are many factors tending to modify the extent of this saving and therefore, it will be well to be conservative in making up estimates of cost based on the theoretical saving to be obtained.

CHAPTER IX

THE USE OF HEATING DATA IN MAKING ESTIMATES ON THE COMPARATIVE COSTS OF ISOLATED PLANT AND CENTRAL-STATION SERVICE

THE preceding chapter reviewed the relation between steam and electrical load in various types of city buildings. In the sixth and seventh chapters, a number of rules were given for estimating the heating requirements in various types of buildings. Next in order is to take up the problem of making a thorough analysis of heating and power plant costs in any given building.

The use of central-station service for the small consumer is almost a foregone conclusion. In large buildings and factories the use of isolated-plant service was formerly the universal rule. The rapid reduction of central-station rates for electricity for the use of the large consumers and a more thorough appreciation on the part of the public of the many advantages of central-station service have conspired to bring about a great change of sentiment among building owners in the last few years. Some of the largest metropolitan buildings are now buying both electric and heating service from the central-station companies, and in every large enterprise which is projected, the question of the advisability of manufacturing the electrical service by a private plant, is a much mooted subject.

Every owner wishes to figure out for himself, if possible, the solution of the problem and in order to make a satisfactory estimate it is necessary to make a very careful and intelligent analysis of the operating conditions of the particular building to be considered. The first step in this analysis is to form an estimate of the annual requirements of the building, for light, power and heat, in other words, to determine the load curves.

These estimates should be made, by first making a complete plan and schedule of the light and power required, and the probable consumption of electricity in every section of the building. After this estimate is made up very carefully, it should be checked with the consumption of electricity in a number of similar buildings, and if the estimate proves to be radically different from the average result found in other buildings the figures should be carefully looked over as they probably contain some error.

It has been found in collecting data from various large city buildings that there is quite a strong similarity in the electrical requirements of buildings in the same business and this material forms a very useful guide in making up estimates. In like manner the steam consumption for heating and miscellaneous steam requirements should be carefully estimated in accordance with the instructions found in the foregoing chapters and these estimates also checked with the known results in similar buildings. These estimates on steam and electrical requirements form the basis of the comparative estimate as to the cost of operation, whether from central station or from a private plant.

By applying the central-station schedules to the estimated requirements, both for steam and electricity, the cost of steam and electricity is very easily computed. Where steam service is not available the cost of operating an independent steam-heating plant can be figured and then added to the cost of the electrical service, based on the schedules of the local company. To these figures should be added the wages of a few men who would be required to oversee the operation of the elevators and other power and to look after the lighting system. Having arrived at the total cost of the service on the central-station basis, it will next be necessary to figure the cost of isolated plant operation, and these figures should be based on the figures of similar plants already installed.

The costs to be included are:

1. Annual fixed charges which are based on the cost of the boilers, furnaces, engines, dynamos, piping and building space required for the installation of the plant.

2. Operating costs which include the items necessary for the operation of the plant after it is installed.

Under fixed charges, we have:

(a) Amortization, which means the amount of money which needs to be set aside each year in order to replace the plant at the expiration of its usefulness. This is usually figured at about 3% of the first investment.

(b) Obsolescence, which is a term used to express the loss in value due to change of conditions. The more important of these changes are changes in the business itself, requiring different types or sizes of apparatus than that originally installed. 2. Improvement in the types of apparatus manufactured, thereby making old apparatus obsolete and unsatisfactory. 3. Reduction of rates by the central-station companies, making the investment unprofitable long before the machinery is worn out. The item of obsolescence is variously estimated, but an average estimate would be about 5%.

(c) Interest. Interest rates vary from time to time, but considering the risk of a plant installation, 6% may be considered as a fair rate of interest.

(d) Repairs. The item of repairs is often included in operating costs but as it varies so much from year to year and is usually higher as the plant grows older, in order to get an average cost of repairs it is simpler to take a fixed percentage of the cost of the plant as the average cost of repairs and this is usually placed at 2%.

(e) Taxes. Taxes vary in different localities. In cities like Chicago, the item of taxes runs from one to one and one-half per cent, but an average would probably be one per cent.

(f) Another fixed charge which can be estimated in two ways is the rental value of space, (1) as the value which could be secured by renting the space to the tenant or (2) it can be figured on the basis of the interest and depreciation on the building investment necessary for taking care of the isolated plant. Often in large cities where space is valuable, very expensive excavations are made in order to provide space for the independent plant. If the annual fixed charges on these excavations is added to the operating cost, it is found that the plant has been an expensive luxury.

(*g*) To the above charges should be added, what is sometimes called " the marginal cost " of operating a plant. This cost is estimated on the following basis:

Almost every large business is limited by its ability to secure capital to carry on the business. If a business is a profitable one, there is usually a considerable margin over the actual cost of borrowing money—perhaps 15% would be the average turnover on the actual money invested in a private business. If money is used in the installation of a private plant, it means so much capital deducted from the business itself, or else it means a straining of the credit of the organization. It is very frequently found that owners will invest large sums of money in private plants only to find that they have made unprofitable investments and at the same time have taken money from their business which is very much needed. While it might not be fair to charge the entire difference between the cost of money and the average annual profit, yet the marginal charge is without doubt a reasonable one and an average estimate might be placed at 5%.

OPERATING COSTS

The operating in connection with a private plant may be sub-divided into two main divisions, viz., Salaries and Supplies.

Salaries include wages paid to engineers, boiler-washers, plumbers, steam-fitters, electricians, oilers, firemen, coal-passers, engineer's clerk and such office help as is required for looking after the operation of the plant. The time of the manager of the organization or owner which would be taken up in the supervision of a power-plant organization should also be included.

Supplies include coal, transportation of ashes, oil, waste, water, shovels, fire-tools, electricity for light, power and ventilation, and handling of coal in boiler and engine-rooms, lamps, carbons, miscellaneous supplies.

The item of Insurance should be divided into Employer's

liability insurance, which pertains to salaries, and boiler and fire insurance, which pertains to supplies.

These items may be given in tabulated form as follows:

1. *Fixed charges based on investment in*

Building,	Dynamos,
Boilers,	Furnaces,
Piping,	Engines,

and various accessories for the above.

(*a*) Amortization.
(*b*) Obsolescence.
(*c*) Interest.
(*d*) Repairs.
(*e*) Taxes.
(*f*) Rental value of space.
(*g*) Marginal charge for diversion of capital.

2. *Operating Costs.*

Salaries.
- Chief Engineer.
- Assistant Engineers.
- Firemen.
- Coal-passers.
- Oilers.
- Electricians.
- Steam-fitters.
- Boiler-washers.
- Elevator repair men.
- Helpers.
- Engineer's clerk.
- Office labor for metering and billing.
- Employer's liability insurance and salaries paid to injured employees when off duty.

Also a portion of the time of the management used in buying supplies and looking after the operating organization.

Supplies.
- Fuel.
- Transportation of ashes.
- Oil, waste, water.
- Shovels, fire-tools.
- Electricity for lighting and power in boiler-room.
- Boiler and fire insurance.
- Miscellaneous supplies and expenses.

All of the above are direct costs which are directly chargeable to the cost of operating a plant. In addition to the above costs, there are other costs which might be termed indirect charges which often come as a result of power-plant operation.

1. Throw over switch service from central-station service. As a rule the cost per K.W. hour for throw-over switch service is greater than the rate where complete service is furnished, and often a minimum bill is required in addition to the higher rate.

2. Danger of breakdown in the service and consequent loss if throw-over switch is not installed.

3. Losses on account of decreased rental value of the building. The majority of isolated plants operated with high-speed engines shows a marked fluctuating quality in the light. This is usually increased at irregular intervals where high-speed electric elevators are operating on the same plant. It is also frequently found that in the summertime the space directly above the boiler is hard to rent on account of the heat coming up through the floor from the engine-room below. There is also the damage and annoyance caused by vibration in the building.

4. Losses of time on account of obstruction of entrances by coal teams. Some of the firms which have discontinued the use of their own plants and gone on central-station service have been particularly desirous of getting the steam service also in order that they might discontinue the delivery of coal to their buildings, and thereby be able to receive and deliver goods without any interference with coal teams.

5. Losses on account of smoke fines, or dirt in the building, due to operation of the boilers. In large western cities where soft coal is used, there has been an active campaign started to prevent the emission of smoke, and a number of these cities have laws imposing fines on the owners of smoky chimneys.

6. Losses on account of strikes, due to labor troubles.

If the above costs of operation are carefully tabulated and are based not on the theoretical economy of apparatus operating at maximum load when new and under special conditions, but on the average operating economy as found in plants, they will show a substantial saving by the use of central-station service providing the rates for central-station service correspond with those recently made in many of our large cities. However, it is often easy to prepare figures which may appear theoretically correct, which give an entirely different result and perhaps show a saving by the use of an isolated plant. These figures fail to take account of the fact that where human agencies are employed there are always some mistakes being made and more or less waste from one cause or another.

Very few isolated plants can afford to purchase a high quality of talent in the operation of their machinery, and the result is that many things are continually going on in the engine-room which militate to a great extent against economy of operation. The moment a plant is installed, no matter how well it is equipped, is the moment it starts to decay. If it is constantly oiled and kept in operation, it may appear to wear out, but there are various chemical causes of deterioration which are constantly at work, as for example, oxygen forming rust, formation of scale by various deposits, and various changes in the substance of which the apparatus is made.

In other words, no matter how carefully a plant is installed, it is only a few months before a great many things require attention and repair. It is not profitable to keep up a plant to the high efficiency which it has when new. The plant owner would spend most of his time in making replacements.

In figuring the average cost of operating the plant, there are always losses to be figured based on the average efficiency of a partially worn-out apparatus. The ordinary engineer in making up figures is apt to figure his coal on the theoretical B.T.U. contained in a pound of coal and he forgets the losses which come through poor coal occasionally delivered by the coal companies, the loss caused by excessive moisture during rainy days in the winter time, and other losses caused by frozen coal. All these contingencies must be met from day to day in the operation of every power plant and these combined losses have a very important bearing on the general result at the end of the year.

It is therefore necessary for the central-station salesman to have as much actual data as possible on the cost of operation of various plants of known capacity so as to give the reasons for these various allowances which must be made in figuring out the cost of operation.

Need of Diplomacy. All the above information and data, however, will amount to nothing, unless a salesman has well in mind the bearing which the different interests related to the proposition hold. It is an old saying that " Convince a woman against her will, and she is of the same opinion still," and while we do not hear this proverb applied to the men, it is apt to be equally true in the case of the average person. It is therefore necessary in presenting the claims of the central-station service to look the ground over carefully and see what the various interests are, which have to do with the problem and the best way in which they should be approached. In some propositions, there is no one but the owner to deal with. In others, there are various parties interested. For example, in the case of the large office building, there are the following parties interested:

1. The central station, which is always interested in the sale of electricity.

2. Manufacturing concerns, whose interests are in the line of selling apparatus not only for the use of electricity from central-station service but also for the production of electricity by isolated plants.

3. The consulting engineer, whose work is not only to design a distribution system for heat and electricity in the building, but also to design generating plants for the production of heat and electricity.

4. The operating engineer, who is often an old and trusted employee who is consulted by the owner as to the best methods of arranging and operating the engineering features of his building.

5. The building manager, who is keenly interested in the question as to whether an isolated plant is installed or not, as it is to his interest to make the best showing possible at the end of the year in the operation of the building.

6. The architect of the building, who is usually closest to

the owner and whose advice as to the net result of the building up of the property has probably induced the owner to make the investment.

7. And lastly, there is the owner, who is of course anxious to secure the best returns from his investment.

Sometimes the central-station salesman makes the mistake of feeling that the six intervening interests which we have just mentioned, are all opposed to the use of central-station service. It is very frequently the case that managers of central stations are inclined to think that they are not given fair consideration as to the use of central-station service, due to the fact that large new buildings are usually designed by architects and engineers whose pecuniary interests appear to be in the line of the installation of a power plant and its accessories, inasmuch as they receive a commission on the apparatus installed. It is true that in a great many cases, the owner looks primarily to his architect for advice as to the various details, not only in constructing but also in operating the building and therefore the best method of obtaining heat, light and power for the property.

It is very unfair, however, to assume that the architect is necessarily inclined to install an expensive power plant simply to enhance his commission. If this same idea were followed to its limit, the architect would also include the most expensive type of construction and would recommend that the work be let to the highest bidder. On the contrary, all first-class architects consider it a part of their duty to see that the owner secures as far as possible the greatest return for the least investment.

There is undoubtedly keen competition among architects as well as in other professions and the architect who can show first-class buildings erected at the lowest cost per cubic foot, will undoubtedly receive consideration from the prospective client. There is, therefore, a very strong inducement for the architect to keep out all apparatus which is unnecessary and everything which may be considered as " frills " in the construction of the building, in order that he may show future clients that he has made a remunerative investment out of the work entrusted to him by his former customers.

Frequently an architect will be engaged by some very wealthy man or corporation for putting up a series of buildings, and the inducement for investing in new enterprises is usually a satisfactory result in the old. It should also be borne in mind that an architect's office is the focus of a great many trades and different lines of business. For example, all the manufacturing interests which are engaged in the construction of dynamos, engines, boilers and power-plant accessories will naturally be on hand, with their advice and recommendations in regard to installing the power plant. The solicitor of central-station service not only must contend with all these various trades but also will contend with all the competing representatives of these various trades, who although they are competitors among themselves are a unit in opposing the adoption of central-station service which would thereby cut off the chance of a sale. The architect is therefore often likely to be influenced where a strong argument is put up from so many different directions and the central-station salesman keeping all this in mind must make due allowances for the various points and arguments.

If the salesman will go to the architect fully prepared with a careful analysis of the job, not trying to secure the business on the basis of generalities, but rather on the basis of a carefully considered estimate, he will find in every first-class architect a man who has the keenest interest as to the cost of maintenance, income from rentals, fixed charges, and all the elements that go to make up the profitable operation of a large property. It is natural that the man who has made the original estimates be deeply interested in the question of light, heat and power supply.

The successful architect is always a broad-minded man and his ability to see both sides of a question should commend itself to the central-station solicitor or engineer. Every great architect is a man who has before his mind the idea of a great and beautiful city; and while he is often forced to be economical in the construction of his buildings, he is always looking toward the city beautiful as the highest exponent of modern civilization.

The use of central-station service with its freedom from smoke, dirt, and dust, is a step in this direction. Therefore, it would seem entirely natural that the broad minded architect would be disposed to give a most cordial reception to the central-station proposition, provided it appeals to his mind as an economical source of supply.

It is always more or less of a problem in the case of any given building as to just who is the best person to approach in regard to obtaining the business, and this question can always be best solved after getting a thorough knowledge of the situation and the personality of the men engaged in the different branches of the work.

It may be laid down as a general rule that the architect of the building is the first man to approach and to interest in the central-station proposition. If the matter is taken up in any other way the salesman will frequently antagonize the architect at the outset and perhaps upset all chances for a favorable consideration. If the architect becomes somewhat interested in the matter he will frequently ask the central-station salesman to take the matter up with his consulting engineer and try to find some common ground on which they can estimate the probable cost of the two schemes. He may suggest that the matter be submitted direct to the owner as this is often a matter in which the architect feels a little delicate in regard to giving positive advice.

If the agent or manager of the building has already been selected, it is always well to make his acquaintance as he may be employed on a basis which stipulates that he would have some voice in the expenses of the building.

In taking up the argument of central station versus isolated plant, the salesman should prepare himself especially on all the various factors, which enter into the heating problem. The engineer of a plant will always come back to his first plea, viz., " using exhaust steam for heating." In other words, he will claim that either the heat is secured for nothing or the electric lighting and power is secured for nothing in the winter time, due to the fact that he is using exhaust steam from his engines in heating the building. We have already found in the preceding chapter that this argument is only partially correct.

Let us now look a little more closely into the subject of heating by exhaust steam. It is not in all respects as satisfactory as has been represented. One of the first disadvantages met with is that oil is distributed all through the heating system, wherever exhaust from the engines has been used for a long time. To be sure oil separators can be used which will diminish this to a certain extent, but in nearly every combined heating and power plant there is found not only oil in the general heating system, but also the returns come back to the boilers with more or less oil, causing trouble in the boilers themselves. This, of course, occurs only where the steam passes first through an elevator pump or steam-engine and the cylinder-oil becomes mixed with the steam and circulates through the system. Where the boiler plant is operated for steam heating only, there is of course no occasion to use oil and the system remains intact.

Another disadvantage of the power plant as compared with the straight-heating boiler is the high-pressure it is necessary to carry in the boiler. It is not necessary to explain to any one that the higher the pressure in the boiler, the more danger there is of explosions and the various troubles to which boilers are subjected. The ordinary heating systems are usually operated at a moderate pressure with reducing valves at the various buildings to adjust the pressure to their requirements. Another disadvantage in the use of a power plant as opposed to the ordinary heating system is the excessive heat on the first floors and basement during the summer time. In the ordinary large building where steam is only required for heating and hot-water service, it is simply necessary to operate a small water-heater by means of hard coal or coke during the summer time. This requires about the same attention as the ordinary furnace in a private house.

An isolated plant requires boilers operated at high pressure all summer long, resulting in intense heat in the basement which usually communicates itself to the offices and stores on the first floor. An advantage of central-station heating service is the reliability on account of the large source of supply. Many buildings have simply a single small boiler to supply them with heat and if anything happens to this boiler, they are in trouble. When steam is sold on the meter basis, the consumer can turn the steam on and off as he needs it and pay for only what he uses. There is therefore a great tendency to eliminate waste of steam, whereas in many office buildings, operating a power plant of their own, steam is kept turned on whether it is needed or not and often when it is injurious to the occupants of the building.

In order to illustrate the method of analyzing the comparative costs of operation in a large city office building, the following are figures on a building recently analyzed in the city of Chicago. This building is a large office building about 200 ft. square and twenty-one stories in height.

It has a court in the center above the first floor 73 ft. square. The original estimate of the steam consumption based on the formulæ given in this book was 63,200,000 lbs. of steam. The actual consumption during the year 1913 as shown by meters was in round numbers 64,300,000 or about 1,100,000 lbs. over the estimate.

As the steam consumption in any building will vary ordinarily a much larger percentage from season to season, the estimate given may be considered fairly accurate. The original estimate for consumption of electricity was 1,250,000 K.W. hours. The consumption in 1913 was 1,100,000 K.W. hours. If a plant had been installed in the building, the consumption would probably have been about 50,000 K.W. hours more, and as the building is not quite rented a complete rental of the building would probably bring the current consumption very nearly up to the estimate.

The actual consumption for this building was, in round numbers 500,000 K.W. hours for tenants lighting, 150,000 K.W hours for public lighting and 450,000 K.W. hours for power, of which about three-quarters was consumed by the elevator equipment. Assuming a price for electricity of $2\frac{1}{2}$¢ per K.W. hour, from the central-station service and a price of 40¢ per thousand pounds for steam on central-station service, it is very easy to figure the cost of central-station service on this basis.

Let us assume that the building will be fully rented and that the total consumption is 1,150,000 K.W. hours. We will also assume that the building purchases its entire requirements both for steam and electricity and retails the electricity to its own tenants. The total bills for the building would be:

1,150,000 K.W. hours at 2½¢ per K.W. hours.....	$28,750.00
64,300,000 lbs. of steam at 40¢ per thousand lbs....	25,720.00
Total...................	$54,470.00

In figuring the cost of isolated-plant service, it will be necessary to add the cost of electricity for lights in engine and boiler-rooms, and also the cost for ventilating same. Assuming, therefore, that this amounts to 50,000 K.W. hours per year, the total electricity used by the plant would be 1,150,000 K.W. hours plus 50,000 K.W. hours or 1,200,000 K.W. hours per year. The average steam consumption in office building plants as shown by a number of tests taken on typical installations is about 60 lbs. of steam per K.W. hour throughout the year. While the above would represent average conditions, in this comparison it would be better to assume 50 lbs. since in a large building such as this, it would be possible to get an economy above the average.

1,200,000 K.W. hours of electricity at 50 lbs. per K.W. hour would require 60,000,000 lbs. of steam per year. From the discussion and curves in the preceding chapter it would be fair to assume that about 40% of this would be saved for heating by utilizing the exhaust from the engines. This would leave a net steam consumption of 60% of 60,000,000 or 36,000,000 lbs. It has been shown by meter readings that the heating requirements of the buildings are 64,300,000 lbs. of steam. Adding together the steam required for electricity and the steam for heating, gives a total of 100,300,000 lbs. or in round numbers 100,000,000 lbs. of steam per annum. The average evaporation in this plant runs about 5 lbs. of steam per pound of coal. If a power plant were operated all summer long the average evaporation would be somewhat higher, say 5½ lbs. of steam per pound of coal. On the basis of 100,000,000 lbs. of steam, the annual coal consumption would be 18,181,818 lbs., or in round figures 9,000 tons. On this basis the operating expenses would be as follows:

Supplies

9,000 tons of coal at $2.75 per ton...........	$24,750.00
Ash removal—6%.......................	1,485.00
Water—for steam supply, washing out boiler and engine-room, etc....................	1,000.00
Oil, waste and packing.....................	1,200.00
Tools and miscellaneous supplies........	1,200.00
Boiler and fire insurance......	60.00
Total	$29,695.00

Labor

Chief Engineer...................	$3,000.00	
Assistant to Chief Engineer........	1,500.00	
3 Watch Engineers..	3,600.00	
2 Oilers........................	1,920.00	
Engineer's clerk......	480.00	
3 Firemen at $840.00	2,520.00	
2 Ashmen at $720.00	1,440.00	
Liability insurance and losses from sickness among employees.......	1,000.00	
Time of office, including manager's time for supervising	1,000.00	$16,460.00
Total operating expenses.........		$46,155.00

In addition to the operating costs we must include the:

Fixed Charges. To take care of this building which has an aggregate installation of about 15,000—50 watt lamps, 200 H.P. in general power and 600 H.P. in elevator power, or a total connected equipment of about 1800 H.P., it will be necessary to install a plant of about 1200 K.W. which would cost complete at $50.00 per K.W. about $60,000.00. The plant would also require space of upwards of 6000 sq.ft. On the above basis, the fixed charges would be as follows:

Amortization at 3%...................... $1,800.00
Obsolescence at 5%....................... 3,000.00
Interest at 6%........................... 3,600.00
Repairs at 2%............................ 1,200.00
Taxes at 1%............................. 600.00
Rental value of space at 50¢ per sq.ft.......... 3,000.00
Marginal charge for diversion of capital at 5%. 3,000.00

Total.................... $16,200.00

Summarizing the above, we have:

Operating Charges........................ $46,155.00
Fixed Charges............................ 16,200.00

Total.................... $62,355.00

It will be noted that the cost for labor to take care of the elevators, electric fans, etc., as well as the radiation has been omitted from both estimates as they are practically equal in both propositions. Comparing this with the above cost of central-station operation, we find a saving of about $8000.00 per year. As a matter of fact the central station costs in Chicago are slightly under these figures. If the price for electricity, however, were 4¢ per K.W. hour and the price of steam 50¢ per thousand pounds, the situation would be reversed, and there would be a saving of about $16,000.00 in the operation of an isolated plant. In other words, the result is not determined by the cost of isolated-plant operation, but by the rates offered by the central-station company.

The above figures are given as average figures and may be found to be higher or lower in different localities and in different plants. The fact that some of the largest buildings in Chicago now operating plants, are running at considerably higher expense than that assumed in this estimate tends to show that the estimated cost of isolated-plant service is conservative.

The above example is given merely as a guide to show the method of analyzing a given proposition and as an illustration of how accurately the consumption of a modern building can often be foretold by a careful study of the conditions beforehand.

The estimates on consumption of electricity were checked at the time they were made by a comparison with the results in similar buildings. The estimates on steam consumption were based on the information given in the preceding chapters. Inasmuch as any estimate of operation would be incomplete without including the cost of the heating service, it is impossible to make an intelligent study and analysis of the requirements of large buildings without thorough investigation and experience in the costs of heating these buildings.

It is hoped that the above discussion may be of some assistance to those who are taking up this problem, whether they are approaching it as representatives of a central station who are endeavoring to obtain large business, or as representatives of real-estate owners or firms of architects who are interested in securing the best results in the most economical manner.

CHAPTER X

RELATION BETWEEN CENTRAL-STATION HEATING AND CENTRAL-STATION LIGHTING AND POWER

It has already been brought out in the first chapter of this book that central-station heating and central-station lighting were started entirely independent of each other and by men who apparently had no conception that they would in any way be supplementary.

Although central-station heating antedates the beginning of central stations for electric lighting and power, its growth has been comparatively slow due to the many technical difficulties to be overcome and the lack of a sufficient financial return on the investment. In view of the unfortunate experiences of many central-station heating companies some have been inclined to discredit the idea of central or district heating and to advocate an independent-heating plant for every building; others, however, have gone on in spite of difficulties and discouragements and established extensive systems for heat distribution. These heating companies have followed two entirely different schemes, the first scheme has been that of using the power station which produces the electrical energy as a source of steam or hot-water supply, to be used in district heating. Examples of these are found in the heating plants of Indianapolis, various heating plants in Milwaukee, the hot-water system of Toledo, Ohio, one of the two main heating stations of the Detroit Edison Company, and the isolated-heating plants operated by the central-station company in St. Louis. A great many other examples might be cited from all over the country and in fact a great majority of the small heating installations are used as adjuncts to electrical-power plants.

Where the central stations for electric light and power are very large, the second scheme for heating is used, viz.: separate steam or hot-water heating plants are installed for the heating only, the electrical power being supplied from large stations which are operated condensing, and usually with steam turbines direct-connected to alternating-current generators. Examples of this second scheme are station No. 2 of the Detroit Edison Company, the New York Steam Company, New York Service Company, some of the heating plants of the Edison Illuminating Company of Boston, and most of the heating plants of the Illinois Maintenance Company in Chicago. Some of the above-mentioned companies originally went into the heating business with the idea of making money on the heating. Others were almost compelled to install heating plants to meet the competition of plants which offered both electric and heating service but none of these companies expect to withdraw from the heating field.

From the standpoint of the electrical power station, it is necessary to furnish heating service:

1. To meet the competition of the isolated plant.

2. To meet the competition of block plants.

Taking up the first cases, it will be found that a great many customers are willing to adopt central-station service throughout if they can be entirely rid of the trouble and annoyance of installing and operating a mechanical plant. They are inclined to argue, however, that if they are required to install a steam-generating plant and hire the men to run it, they might just as well make the plant a complete one and furnish the complete electric and heating service to the building. Whether this view is right or wrong, the fact remains that it is taken by a great many intelligent business men and if the central-station company is to secure the business they must be prepared to also furnish satisfactory steam service.

In March, 1915, the Union Electric Light & Power Company of St. Louis writes as follows:

" We find that in cases where isolated-generating plants are already installed, we are obliged to take over the plant and operate it during the winter months in order to obtain the business; while in the case of new buildings, the builder is installing only the boilers necessary for heating which he operates himself depending on the central-station system entirely for his electrical service."

The experience of the St. Louis Company is especially interesting to central-station men as that company has been eminently successful in shutting down isolated plants, especially during the past two or three years.

The second form of competition is the block-plant. This usually comprises a small power plant which has been installed in some private building and which furnishes not only electricity for the building of the owner, but also makes arrangements to supply the wants of its neighbors in the same block and sometimes in one or two adjoining blocks.

The investment costs of these plants is comparatively small, the transmission lines are very short, reducing the investment in copper and also the drop in potential to a small amount. Formerly these block plants devoted their attention exclusively to obtaining lighting and power customers, but in the past few years most of them have realized the advantages of also including the heating service, using as far as practicable the exhaust steam from their engines for this purpose.

Where the central-station company which has no auxiliary-steam system meets the competition of one of these plants which provides complete service the central-station is usually defeated, as the customer prefers to buy his electricity where he can also buy heating service and thus do away altogether with the necessity of a plant. A great many customers have been lost to the central-station who might have been obtained had it been in a position to furnish the heating service. For this reason many central-stations have awakened to the fact that they too must be prepared to furnish a complete service unless they are to lose a large part of the business which they formerly held.

One of the mistakes which have been frequently made by central-station companies where they have started on this policy is to figure that the heating service should be rendered at less than cost, in order to obtain the contract for electrical service. Many a steam contract has been made for steam service at ridiculously low prices which have caused a great deal of loss to the heating company where it would have been almost as easy to secure an adequate price if the company had started out in the first place with that idea. Some managers seem to be obsessed with a desire to make special concessions whenever they secure a large piece of business and they consider it as evidence of business sagacity to cover it up by means of a flat rate on the steam heating. Without attempting to argue the point as to whether it may or may not have been advisable in the past to make special concessions on heating contracts, the fact remains that at the present time the day when such transactions were advisable if they ever were advisable has gone by.

The heating contract should stand fairly and squarely upon its merits and demand an adequate " quid pro quo." Public sentiment has been aroused to a very high standard in dealing with public-utility corporations and the demand now is for a a fair price to every one and no favors. The establishment of state commissions for investigating and regulating the rates of public-service corporations and the drastic laws which have been enacted to enforce these regulations are sufficient proof of the present trend of public opinion. The public does not want to have steam heat in order to get something for nothing but demands heating service on the broad basis of convenience and economy. The advantages to the public of district-heating service are numerous, and among them may be mentioned:

ADVANTAGES TO PUBLIC OF DISTRICT-HEATING SERVICE

1. **Reduction of Labor.** This would include not only the labor involved in the operation of the steam-generating plant, but the various repairs, supervision, etc., required in connection therewith.

2. **Reduction of Cartage Through City Streets.** In every great city of the country, one of the most vital problems is that of street congestion. Quite a large portion of the cartage

through city streets now consists of coal wagons, conveying thousands of tons of coal every day to the different buildings and plants of the city. By the adoption of district-heating stations the coal could be supplied either from railway cars direct to stations, or from nearby points, reducing the hauling of coal to a minimum.

3. **Abolition of the Smoke Nuisance.** A few years ago, it was supposed to be a natural necessity in all large towns in the central part of this country that there be a continual stream of smoke pouring from the chimneys of manufacturing establishments or large buildings, filling the air with soot and defiling practically every object of the city. As explained in preceding chapters, this matter has been made in recent years a subject of scientific study and it has been found possible, especially in the larger and better equipped stations to reduce the production of smoke to a very small fraction of that heretofore supposed necessary. It is needless to say that the centralization of heating supply and consequent production of heating service from large and well equipped plants will be a great step forward in the cleansing of the city atmosphere.

4. **The Item of Cleanliness Around the Premises.** Even those plants which are operating without smoke have to contend to a greater or less extent with clouds of dust which arise when coal is delivered to the premises. These clouds of dust spread all over the sidewalk, permeate the engine-room and boiler-room and diffuse themselves out through the various portions of the building. Steam or hot-water service is delivered in pipes which are absolutely clean and wholesome.

5. **Safety.** Any one who takes the pains to keep posted on the matter knows that boiler explosions are constantly occurring all over the country. While it is hard to tell the cause in many cases, the fact remains that they usually occur in the cheaper and more poorly-equipped plants. The more destructive explosions usually occur with the cheaper types of tubular boilers The centralization of steam supply will naturally cut down the number of boilers and also tend to the production of steam from the larger and safer types of water-tube boilers.

The distribution of heat from central-stations whether it be by means of steam, hot water, or by low-priced gas, although it may be considered a luxury today, will undoubtedly be viewed as a necessity in the near future. The public demands this service and what the public demands some one will have to furnish. As has already been shown, the great central-power companies will be forced into giving this service by the pressure of competition whether they wish to enter into the business or not. It is only a question of time when the successful and up-to-date stations will be furnishing this service and if they do not they will be crowded out by the competition of others who will do so.

In this connection, it may be of interest to cite a few examples of progressive central-stations which have already taken a serious interest in district heating, in addition to those examples already enumerated in Chapter I.

The accompanying map (Fig. 79), shows five of the six heating stations of the Edison Electric Illuminating Company of Boston. The connected radiation, both direct and indirect is equivalent to about 150,000 sq.ft. of direct-steam radiation. When to this is added the steam supplied to restaurants, hydraulic elevators, kitchens, etc., it will total a connected-load equivalent to about 200,000 sq.ft. of radiation. Mr. R.S. Hale of this company who is also chairman of the rate committee of the National Electric Light Association, has expressed himself in favor of a fair and equitable steam-rate commensurate with the service furnished.

Across the continent lies San Francisco, a city which has paid a great deal of attention to electric illumination and power. In this city also the subject of steam heating has received considerable attention. The accompanying map (Fig. 80), shows the steam-heating system of the Pacific Gas & Electric Company. This company supplies steam not only for heating purposes but also in some cases for cooking, and for the operation of pumps. The equivalent connected load is upwards of 300,000 sq.ft. of radiation. Mr. Varney, the chief engineer of operation and maintenance for the Pacific Gas & Electric Company holds very positive views on the subject of steam heating in connection with central-station service. In a letter written January 23, 1915, he states:

" To my mind the steam-heating business should be handled absolutely independent of the electric light and power business. It is my personal opinion that in most cities there is a field for central-station heating-plants, but they should be made to pay their own expenses and not be used as " trading stamps " for the electric business."

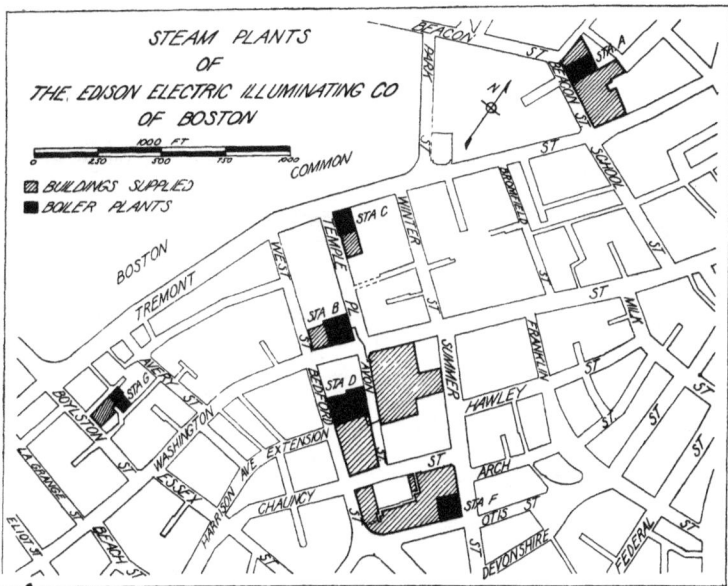

FIG. 79.—Steam plants of the Edison Illuminating Co., in Boston, Mass.

Further on he states:

" In clubs, cafés and some office buildings the steam supplied from the central stations often removes obstacles by inducing the manager to dispense with his isolated plant, as it has been our experience, that the average managers of such establishments are very desirous of eliminating as many of the engineering features as possible from their control. In that way the supplying of steam not only allows them to dispense with the engines and generators, but also with the boiler plant."

Nine hundred miles to the northward is the city of Seattle. The accompanying map (Fig. 81), shows the underground distributing mains of the Puget Sound Traction, Light and Power Company.

In the winter of 1899 and 1900, the American District Steam Company installed for the Seattle Electric Company, now known

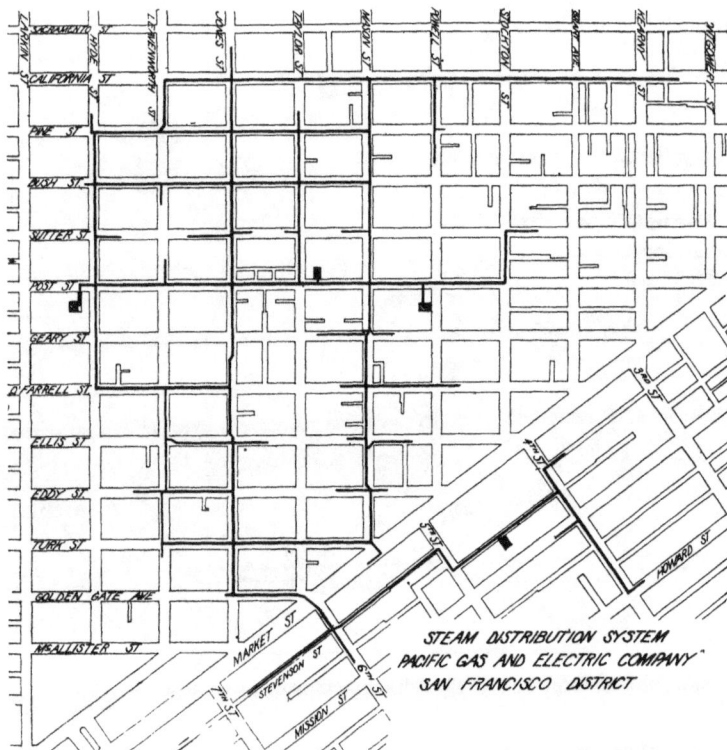

FIG. 80.—Steam-distribution system of the Pacific Gas and Electric Co., in San Francisco

as the Puget Sound Traction, Light and Power Company, an underground-heating system of approximately 6,250 lin.ft. of trunk lines. A few years later the Diamond Ice Company of Seattle, installed underground-heating mains in a district adjoining the district served by the Puget Sound Traction, Light and Power Company. This system has since been acquired by the latter company. Due to the rapid growth of the city and the popularity of the district steam-heating service, the Puget Sound

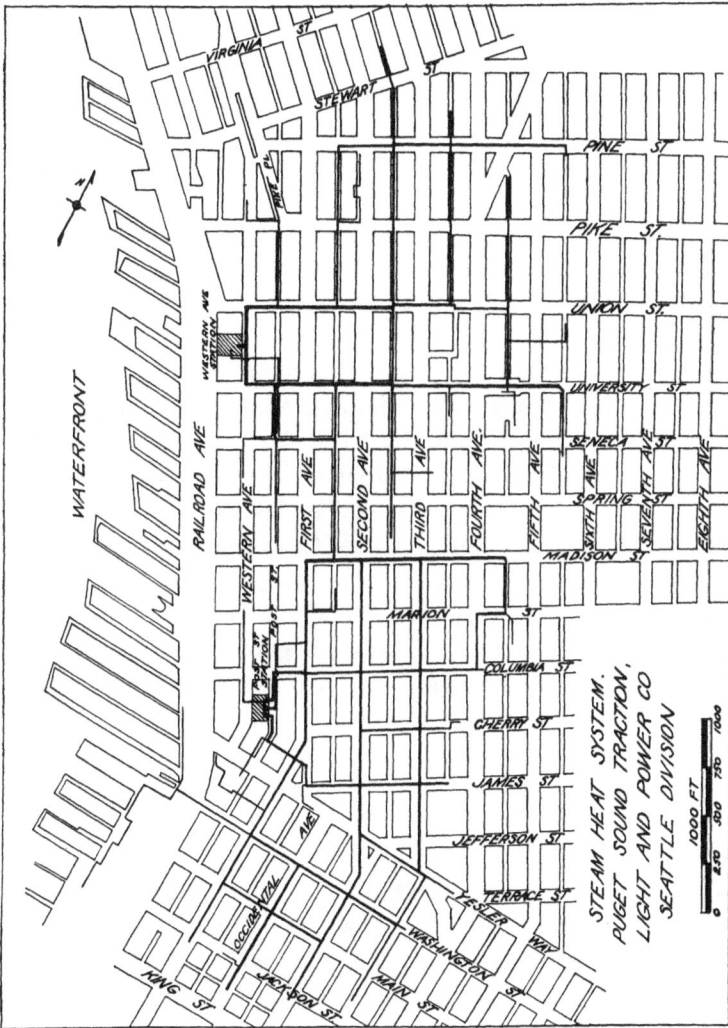

FIG. 81.—Map of the steam-heating system of the Puget Sound Traction, Light and Power Co., in Seattle, Wash.

Light and Power Company has been called upon to make extensive additions to their underground system and in the spring of 1915 they had in operation 32,250 lin.ft. of trunk-lines, and approximately 10,000 ft. of service lines.

At the present time the company has connected about 550 customers and supplies heat for about 150,000,000 cu.ft. of space. Assuming one square foot of radiation to each 125 cu.ft. of space the service supplies upwards of 1,200,000 sq.ft. of radiation. Although the steam-heating business is carried on by the light and power company, it has been found cheaper to furnish the light and power from their hydro-electric plants, the heating service only being supplied from their local steam-plants. In these local steam-plants are also installed dynamos which are used simply as reserve capacity to be started up in case of any failure of the hydraulic system. Although these steam-plants have been operated entirely independent of lighting, the service being supplied by live steam, the balance sheet for the past year shows a substantial net profit in their operation.

The accompanying map (Fig. 82), shows the underground hot-water heating system of the Toledo Railways & Light Company in Toledo, Ohio. This system which covers a territory of about two square miles has already connected about 1,250,000 sq.ft. of hot-water radiation. Mr. A. C. Rogers of this company, who was the first president of the National District Heating Association, and has been for a number of years identified with heating interests, expresses himself as extremely optimistic on the future of district heating. In a letter of February 15, 1915, he states as follows:

" Can district heating be made profitable? Yes, emphatically yes, and to do this it will not be necessary to put rates up beyond reasonableness, but it will be necessary to put in force stringent supervision of both plant and distribution system, and also consumers' buildings, heating installations and operation."

" There is no doubt in my mind that district heating is of great assistance to the central stations in securing electrical business. District heating is fundamentally economical to the business block and is universally sought by all wide-awake concerns. The trend of the times towards conservation and all other economies, the growing public sentiment for " City Beautiful " effects, cleanliness and artistic surroundings will enhance and bring forward the central-power plant as against the individual one. The expansion of the cities into *quasi* rural groups, the platting of additions, with studied effects of architecture and landscape-gardening has received a material and quickening effect from the utilization of district heating."

In the city of Rochester, New York, the Rochester Railway & Light Company, which furnishes the bulk of the power and

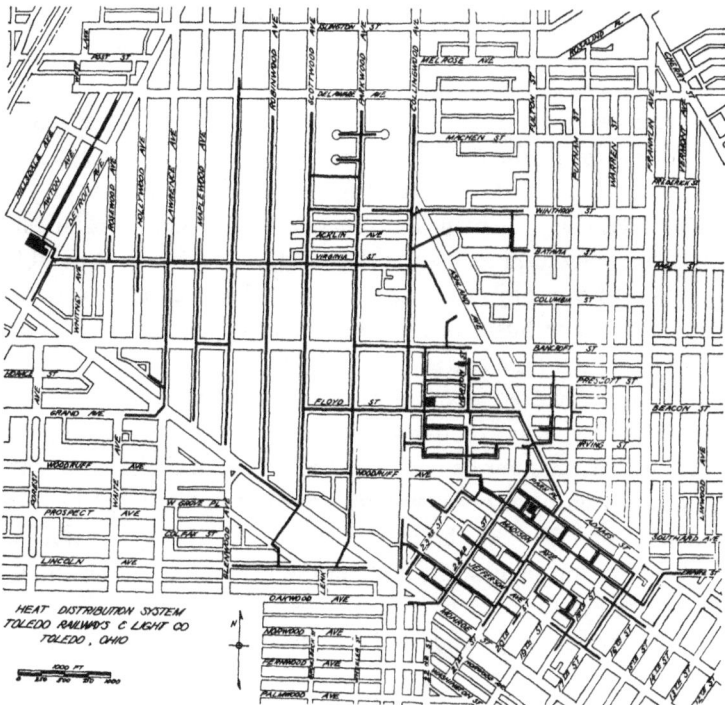

FIG. 82.—Heat-distribution system of the Toledo Railways & Light Company, Toledo, Ohio

light throughout the city has also taken an active interest in the subject of district heating and now has an underground system of district-heating mains with a connected load equivalent to about 280,000 sq.ft. of radiation.

Mr. Arthur Williams, General Inspector of the New York Edison Company, states, in a letter of February 8, 1915:

" Our experience has been that it is not necessary to make

concessions below cost to obtain contracts for heating service. The advantage of having a heating company available is that it prevents an inflation of the architects or engineers estimate upon the cost of heating, in that the electrical company has available a company which, for a stated amount, will undertake the heating of a building should the owner so desire."

" We have an instance here (of many) in which the engineers estimated that the cost of heating a Fourth Avenue structure would be $22,000 yearly. Our own estimate was about $9,000 and we offered to take the contract over at any time for $12,000. The offer has never been accepted, but it was the ability to make the offer through the New York Service Company that enabled this company to get the contract for electric light and power."

Mr. Reginald Pelham Bolton, one of the most prominent consulting engineers in New York City, has given considerable attention to the subject of the connection between the electrical light and power business and district heating. In a letter of February 9, 1915, he states:

" I am convinced that the course of modern economics will lead to the eventual production of heat from central plants in just the same manner as has been the course of developmant in the production of electricity, and I even go so far as to believe that the necessities of the future will demand the cessation of small producing plants of the domestic character, on account of the wastage of fuel which accompanies their operation.

" As the matter stands today, the production of a unit of electric energy requires the dissipation of five times the amount of fuel in a private plant than the rate required in a modern central station. As fuel becomes more expensive, it is evident that the trend of public necessities will be in the direction I have stated.

" The same course is evident in connection with the production of heat in crowded communities. The process of individual heat production is one of extraordinary wastefulness, and this being the case, it appears to me that the interests of the production of electric energy and that of production of heat lie in the same direction, and that it is bound to become the policy of the electric service corporations to supplement their service by that of heating. In our great cities this would not only be good business policy but would serve the public indirectly by the elimination of unnecessary smoke, the reduction of the total consumption of fuel, and the corresponding reduction of dust and relief of traffic in the streets."

Mr. S. B. Way, Vice-President and General Manager of the Milwaukee Railway & Light Company of Milwaukee, has had quite an extensive experience with the furnishing of central-station service for light and power in connection with steam heating in the city of Milwaukee, Wisconsin. Mr. Way states:

" I am heartily in accord with all plans for securing central-station business which do not involve unjust discriminations in favor of certain classes of customers, and I am particularly in favor of the idea that central-station companies should place themselves in position to render a complete service to the customer so as to relieve the customer of all worry in connection with his supply of power and heating service; this, in the end, causes customers to become more and more dependent upon the central-station company and to have a greater degree of confidence in the central-station idea."

" I see no objection, and, in fact, I see considerable advantage to be gained by a policy of operating isolated-plant boilers at low-pressure for supplying heat during the winter months, and it is by this method largely that the Union Electric Light & Power Company in St. Louis have been able to secure a very large volume of downtown office building business. The operation of boilers in individual buildings avoids the heavy investment necessary to install a central-heating system, which is the only other method by which the central station can render full service to customers."

" In Milwaukee, we have one of the most extensive central-heating systems in the country, which makes it unnecessary for us to consider the operation of local boiler plants or isolated plants, except in districts outside of that served by the heating system. So far, we have found no cases where we could secure the business where it was necessary for the central-station company to take over the operation of the customer's isolated plant for certain periods of the year, or to undertake the operation of the customer's boilers to supply heating service."

Many others might be quoted and examples shown of large central stations which have begun to appreciate the importance of combining the production of light and power with the production of heat, but enough has already been given to illustrate the general trend in this direction.

In view of these facts it behooves the electrical fraternity to study carefully the whole question of heat supply and strive to discover economies which will make this a profitable portion of their business.

The great steam-boilers and turbines of central-power stations will produce a kilo-watt hour for the expenditure of a little more than one-fifth the amount of coal required in the central stations of twenty-five years ago, while the small installations have made comparatively little progress. In like manner the future may disclose corresponding economies that can be produced by the centralization of heat supply, thus making it a valued and profitable adjunct in the work of the public utility company.

www.ingramcontent.com/pod-product-compliance
Lightning Source LLC
Chambersburg PA
CBHW031806190326
41518CB00006B/215